SYSTEMATIC BOTANY MONOGRAPHS

VOLUME 44

Systematics of Columnea Section Pentadenia and Section Stygnanthe (Gesneriaceae)

James F. Smith

THE AMERICAN SOCIETY OF PLANT TAXONOMISTS
30 September 1994

SYSTEMATIC BOTANY MONOGRAPHS
ISSN 0373-8211

ISBN 0-912861-44-4

Printed in the United States of America

Publication of volume 44 of Systematic Botany Monographs was partly funded by Walter L. Meagher, whose generous support is gratefully acknowledged.

SYSTEMATICS OF COLUMNEA SECTION PENTADENIA AND SECTION STYGNANTHE (GESNERIACEAE)

James F. Smith
Department of Biology
Boise State University
Boise, Idaho 83725

Abstract. The 27 species of *Columnea* (Gesneriaceae) here described are assigned to sections *Pentadenia* and *Stygnanthe*. They are herbaceous perennials, occurring from southern Mexico to Bolivia from near sea level to 3800 m. In addition to morphology, geographic distribution, breeding systems, and phylogeny are discussed. The taxonomy is based on cladistic analyses of morphological diversity and chloroplast DNA restriction site variation (details of the analyses are published elsewhere). Section *Pentadenia*, previously monotypic, now comprises 9 species of suffrutescent robust herbs with isophyllous to slightly anisophyllous leaves, long pedicels, and conspicuous, strongly ventricose corollas. Section *Stygnanthe*, also newly circumscribed, includes 18 species of creeping, pendent, or upright herbs with isophyllous to anisophyllous leaves, short pedicels, and generally small, inconspicuous, and only slightly ventricose corollas. Three new combinations are proposed in sect. *Stygnanthe*: *Columnea antiocana*, *C. fritschii*, and *C. rileyi*.

"One of the most beautiful plants I have seen in Venezuela."
J. A. Steyermark regarding *Columnea strigosa*

INTRODUCTION

The aesthetic qualities of the Gesneriaceae, particularly *Columnea*, have contributed to the pleasure and excitement in studying this group; however, the true scientific merits of the genus, and family, encompass qualities beyond the physical beauty of the plants. *Columnea* is a diverse genus in terms of morphology and habitat. Although all species are herbaceous, they range from suffrutescent terrestrial herbs to thin-stemmed creeping epiphytes. Various species can be found from wet forests at sea-level, mid-elevation seasonally dry forests, and sub-paramo forests at 3800 meters in elevation. This diversity has provided an intriguing source of data for evolutionary studies (Givnish 1984; Smith & Sytsma 1994c). This same diversity has also resulted in a highly confused and complicated classification. It is the aim of this paper to begin clarifying the classification of *Columnea* through a revision of two of the smaller sections within the genus, *Pentadenia* and *Stygnanthe*.

Columnea is the largest neotropical genus in the Gesneriaceae and comprises over 200 species. The species occur from Mexico to Bolivia and eastward to Amapá in Brazil and in the Caribbean islands. Although there is a great deal of intrageneric variation, the genus is distinguished from other members of the Gesneriaceae by tubular corollas that are gibbous at the base, superior ovaries, anthers coherent at the apex in two pairs with longitudinal dehiscence, and berry fruits. Species of sect. *Pentadenia* are distinguished by a robust upright suffruticose habit with isophyllous or slightly anisophyllous leaves, large ventricose corollas, long pedicels, and inflorescences of generally one to three flowers. In contrast, species of sect.

Stygnanthe are epiphytic herbs with a creeping to pendent habit, small, slightly ventricose corollas, short pedicels, and inflorescences of generally three to twelve flowers.

Because the large size of *Columnea* (over 200 species) made it unfeasible to study as a single unit, sections *Pentadenia* and *Stygnanthe* were chosen. These two sections comprise a reasonable number of species for investigation (27), and the lack of any recent revisionary work covering the entirety of the two sections made them a suitable starting point for continuing work within the genus. These two sections, believed to be closely related, previously constituted the genus *Pentadenia* (sensu Wiehler 1973, 1983). The species examined here had not been studied in a systematic way since initial description or were included in floristic works that treated a portion of the species. Here species descriptions have been modified and enhanced based on accumulated collections and field observations, and keys to species and distribution maps have been constructed, which will be the basis for further taxonomic studies within *Columnea*. In addition, it has been the goal of this study to determine the sectional delimitations of *Pentadenia* and *Stygnanthe*.

Prior to this revision, all but one of the species (*Columnea oblongifolia*) have been considered to belong in sect. *Stygnanthe* or the genus *Pentadenia* (sensu Wiehler). Morphologically, the two sections are quite distinct from each other, but were considered similar on the basis of the five-lobed nectary found in species of these two sections. All other species of *Columnea* possess a single bilobed gland or two separate glands. Recent studies (Smith 1991; Kvist & Skog 1993) found that the nectary lobe number is variable, even within an individual, and may frequently consist of a two-lobed gland. The nectary lobe number is therefore a poor character to unite these two sections even though species of sections *Pentadenia* and *Stygnanthe* are the only species within *Columnea* that can have five-lobed nectaries. Therefore, one of the goals of this study has been to determine whether the two sections were more closely related to each other than to any of the other four remaining sections of the genus. Cladistic analyses were performed using chloroplast DNA restriction site variation and morphology. Analyses were performed with the two data sets independently and in combination. The results of these studies are briefly discussed in this paper but are described in detail in additional papers (Smith & Sytsma 1994a, b, c).

TAXONOMIC HISTORY

In the first edition of his *Species plantarum* (1753), Linnaeus listed one species of *Columnea.* Since then over 200 additional species have been described, making a subgeneric classification desirable. Hanstein (1865), Bentham (1876), and Fritsch (1893–1894) divided *Columnea* into sections, following the system established by Hanstein (1865).

Wiehler (1973, 1975, 1983) reexamined *Columnea* s.l. and divided it into four segregate genera: *Dalbergaria* Tussac, *Trichantha* Hook., *Pentadenia* (Planch.) Hanst., and *Columnea.* He also described a fifth genus, *Bucinellina* (Wiehler 1977, 1981). Morley (1976a) challenged Wiehler's (1973, 1983) classification and defined five sections within *Columnea* s.l. Morley (1976a) claimed that the elevation to generic status by Wiehler (1973, 1983) did not create more natural groups than those of the sectional classifications already existing.

Skog (1979), in revising the Gesneriaceae for the Flora of Panama, included all species of *Dalbergaria, Columnea, Pentadenia*, and *Trichantha* in *Columnea* s.l. Kvist and Skog (1993) revised the sectional classification for their treatment of *Columnea* in Ecuador and recognized *Dalbergaria, Columnea, Pentadenia, Trichantha*, and *Bucinellina* as sections of *Columnea* s.l. They also reinstated sect. *Stygnanthe* to comprise the majority of species in Wiehler's (1973, 1983) genus *Pentadenia*. Only *Columnea strigosa* remained in sect. *Pentadenia.*

In the current work, the six sections of Kvist and Skog (1993) [sect. *Collandra* (Lem.) Benth. (=*Dalbergaria*), sect. *Columnea*, sect. *Bucinellina* (Wiehler) L. P. Kvist & L. E. Skog, sect. *Pentadenia* (Planch.) Benth., sect. *Stygnanthe*, and sect. *Ortholoma* Benth. (=*Trichantha*)] are recognized. Several of the species formerly placed in sect. *Stygnanthe* (Kvist & Skog 1993) are transferred to sect. *Pentadenia*. For a more detailed taxonomic history of *Columnea*, consult Kvist and Skog (1993).

MORPHOLOGY

Habit. Habit throughout *Columnea* is highly variable; however, within sect. *Pentadenia* most species are suffruticose, upright herbs, occasionally epiphytic, with opposite, isophyllous to slightly anisophyllous leaves. A few species are scandent shrubs, and only some individuals of *C. strigosa* can be truly classified as vining. Within *Columnea*, the clustering of leaves at the apex of the stem is unique to sect. *Pentadenia*. Clustered leaves are found in species of seasonally dry forests that drop the lowermost leaves during the dry season (Mansfeld 1937; Kvist & Skog 1993).

Section *Stygnanthe* is far more variable in habit, although all of the species are primarily epiphytic or epipetric, growing on open rock surfaces. The plant form ranges from small upright or creeping shrubby herbs to stems that grow horizontally from their base with a dorsiventral leaf arrangement, typical of sect. *Collandra*, to thin-stemmed pendent herbs.

Leaves. Nearly all species of *Columnea* have opposite leaves. In several species one leaf of a pair continues to develop and enlarge, while the growth of the other is arrested (Sánchez-Burgos & Dengler 1988), resulting in very unequally-sized leaves that may even appear to be alternate. Although this trait, known as anisophylly, appears throughout *Columnea* and in other genera of the family, no representatives of sect. *Pentadenia* possess this trait. In addition, the presence of whorled leaves in *Columnea* is unique to *C. fritschii* and *C. ultraviolacea*, one of several characters that unite them as sister species.

Morley (1973a, 1973b) hypothesized that anisophylly may have arisen to improve light capture in an understory herb; leaves could become larger and capture more light without shading each other. Givnish (1984) proposed that while it was possible for leaves to grow larger and remain opposite without self-shading, the plant would necessarily have to produce longer internodes. A stem necessary to maintain larger opposite leaves without self-shading would require higher rates of photosynthesis to produce the longer stems, and the amount of energy required to produce a larger stem would be unfeasible for a light-limited understory herb.

Therefore, anisophylly may have arisen as the combined result of selection for greater light capture and reduced stem volume. Larger leaves can be packed onto the same length of stem without self-shading if they are alternate (Givnish 1984).

Leaf shape is extremely variable within sect. *Stygnanthe*, ranging from small, nearly orbicular laminas in *C. ovatifolia* and *C. colombiana* to long lanceolate-linear laminas in *C. manabiana*. Leaf shape is also variable within sect. *Pentadenia*, although it is not as great as within sect. *Stygnanthe*. Leaves of sect. *Pentadenia* are ovate-elliptic and, in general, are larger than those of sect. *Stygnanthe*.

Leaf vesture is another highly diverse character that can be taxonomically useful if treated as a combination of trichome type, length, distribution pattern, and density on either side of the lamina (Smith & Sytsma 1994b). In general, the pubescence on the abaxial side of the leaf lamina is less dense than on the adaxial surface and usually composed of shorter trichomes. Pubescence on the abaxial surface of midribs and veins tends to be denser and longer than the pubescence on the adaxial side. The pubescence on the veins may be red-purple but not elsewhere on the laminar surfaces. Because the veins of leaves in *Columnea* tend to run to the edge of the lamina, vesture on the veins creates a ciliate margin.

Four different types of trichomes have been identified in *Columnea*: 1) uniseriate sharp-pointed, 2) uniseriate gland-tipped, 3) uniseriate slender, and 4) glandular papillae (Morley 1974; Wiehler 1983). In addition, these trichomes may all be red-purple or colorless. There are also trichomes that are truly multicellular but very short and appear as colorless "single-celled trichomes"; they form a strigose vesture when abundant. Technically, these trichomes are multicellular in that they have a small basal cell that is only infrequently visible even with a dissecting microscope. The longer uniseriate trichomes occur on the veins, whereas the very short multicellular trichomes occur on the lamina. The latter are here referred to as "single-celled" to avoid confusion between short and long multicellular hairs that are clearly visible as uniseriate trichomes. Figure 27B illustrates these very short multicellular ("single-celled") trichomes.

For the most part, leaves bear uniseriate slender, uniseriate sharp-pointed, or the "single-celled" trichomes mentioned above. Glandular trichomes are restricted to corollas.

Leaf coloration patterns are also variable in *Columnea*. Most of this variation is found in sect. *Collandra*, although there are sizeable differences in sect. *Stygnanthe*. The coloration patterns have been proposed as additional pollinator signals in species with less showy corollas (Jones & Rich 1972). Leaves in sect. *Stygnanthe* range from entirely green to entirely purple. This variation is seen even within a species, such as *C. spathulata*. Leaves of sect. *Pentadenia* are less variable in color, and tend to be either all green or with purple adaxially.

Inflorescence. The basic unit of the inflorescence in the Gesneriaceae is the pair-flowered cyme (Weber 1973; Wiehler 1983). In *Columnea*, and in several closely related genera, such as *Alloplectus* Mart. and *Drymonia* Mart., the inflorescence is axillary and reduced to a single-flowered cyme. In *Columnea* several inflorescences of single-flowered cymes may constitute a synflorescence. Although the correct terminology for these structures is synflorescence, an "inflorescence of one to twelve flowers" is used in this text; this designation is more widely used and avoids confusion and excess terminology.

In general, most species of sect. *Stygnanthe* have many single-flowered inflorescences per leaf axil and species of sect. *Pentadenia* have only one or two. All species examined were found to have bracteate inflorescences, although in the majority of species the bracts were either very small and inconspicuous, or caducous.

Pedicel. The pedicel in *Columnea* is variable in length; in sect. *Stygnanthe* it tends to be very short and erect, whereas in sect. *Pentadenia* it tends to be long and pendent. Many species examined have oval to round, dark purple structures on the pedicel at the base of the calyx. The function of these structures is yet undetermined; due to their appearance they are here referred to as glands. These structures have phylogenetic use in that their occurrence is a synapomorphy (albeit homoplastic) for three clades within the two sections examined here (Smith & Sytsma 1994a). They are found throughout the genus and are frequently obscured by vesture or bracts. In some species, particularly *C. strigosa*, they are more prominent in fruit than in flower.

Calyx. The calyx in *Columnea* is lobed nearly to the base. The lobes are generally equal to subequal in size and clasp the base of the corolla. This differs from the condition in *Drymonia* in which the dorsal lobe is smaller and reflexed to accommodate the large gibbous base of the corolla. Also, the calyx lobes in *Columnea* tend to be more lanceolate and narrower than the broader, ovate-elliptic calyx lobes of *Alloplectus* and *Drymonia*. This is an important distinction, because several of the species in sect. *Pentadenia* possess some characters similar to species of *Alloplectus* but have the narrow, lanceolate calyx lobes typical of *Columnea*. Although the main feature used to distinguish *Alloplectus* from *Columnea* is the berry fruit of *Columnea*, the calyx lobes are an additional character that may prove useful when fruits are absent or immature.

In *Columnea*, calyx lobes are highly variable in shape, size, coloration, margin, and pubescence. In general, the coloration, pubescence, and margin of the calyx lobes tend to be the same as those of the leaves. As with most morphological characters, the calyx lobes of sect. *Pentadenia* tend to be larger and broader than those of sect. *Stygnanthe*. The calyx lobe margin of species in sect. *Stygnanthe* is usually entire, whereas it is usually toothed in sect. *Pentadenia*.

In some species of *Columnea* the calyx lobes become enlarged and sometimes more darkly colored in fruit (Wiehler 1977). This trait occurs in both sections and has been detected in *C. orientandina, C. byrsina*, and *C. strigosa*.

Corolla. The corolla of all species of *Columnea* is tubular, gibbous at the base, and has five free apical lobes. In some species the two dorsal lobes are fused. In both sections *Pentadenia* and *Stygnanthe* the corolla is ventricose to some degree. Ventricose corollas are swollen in the middle with a constriction before becoming gibbous at the base, and another constriction before the lobes at the opening (Fig. 3). Most species of the two sections have small lobes; the exceptions are *C. strigosa* and *C. oblongifolia*. Corollas in sect. *Pentadenia* are generally larger and more strongly ventricose than those in sect. *Stygnanthe*, where the less conspicuous corollas have a less pronounced ventricose pouch.

All species of sections *Pentadenia* and *Stygnanthe* (except *C. strigosa*, *C.*

hypocyrtantha, and *C. antiocana*) also have glandular trichomes on the interior dorsal surface of the corolla near the opening. In some species these glandular trichomes extend to the base. The function of the trichomes is uncertain; most likely, they facilitate pollen transfer during selfing. Many species of *Columnea* are self-compatible (Smith 1991), and several species will set fruit in the greenhouse where no pollinators are available (Ertelt 1990; pers obs.).

In addition, the corolla trichomes may be a means to insure that outcrossed pollen comes in contact with the stigma. Pollinators, visiting the corolla before the stigma is receptive, may transfer the pollen to the trichomes. The pollen is later transferred to the stigma when the style elongates. This scenario would insure outcrossing in the event that pollinator visits are infrequent or irregular.

The corolla shape has traditionally been the most useful character for defining taxonomic groups above and below the level of genus in the Gesneriaceae (Fritsch 1893–1894; Morton 1963; Morley 1976a; Wiehler 1983; Kvist & Skog 1993). Wiehler (1983), in his treatment of the neotropical Gesneriaceae, placed less emphasis on the shape of the corolla in determining generic and subgeneric groups and used instead characters of habit, nectary, and pollen. Most species of *Columnea* are believed to be pollinated by hummingbirds (Stearn 1969; Morley 1966, 1971, 1972, 1973a; Jones & Rich 1972; Wiehler 1983). Because hummingbirds are potentially highly selective of feeding source (Linhart & Feinsinger 1980), a classification emphasizing floral characters may not reflect true relationships among the plant species, but instead may be a classification of their pollinators (Macior 1971; Kiester et al. 1984).

Although reducing the importance of corolla form in the classification of *Columnea* may not be necessary, an increased emphasis of characters other than corolla form can be useful. Section *Collandra* is an excellent example; it may be defined by large, strongly anisophyllous leaves (Wiehler 1983; Kvist & Skog 1993).

Androecium. The androecium consists of 4 stamens adnate to the base of the corolla from 1 to 5 mm and connate for an additional 0.7–9 mm. Stamens are mostly included within the corolla, but are exserted in *C. oblongifolia, C. orientandina*, and *C. byrsina*. Filaments may be white, red, yellow, or pink-purple and range from glabrous to slightly pubescent or slightly pilose. Anthers are quadrate, subquadrate, or rectangular.

Pollen. Pollen was examined by Fritze and Williams (1988) and Williams (1978), following the taxonomic system of Wiehler (1973), which divided *Columnea* into five segregate genera. In the pollen analyses of Fritze and Williams (1988) and Williams (1978), both sections *Pentadenia* and *Stygnanthe* are treated as a single genus, *Pentadenia*. Although the pollen surface morphology in *Columnea* s.str. (sect. *Columnea*) was found to be highly uniform, variation in the other genera did not provide substantial evidence to support the generic classification of Wiehler (1973, 1983).

A high degree of uniformity was noted in the pollen among the species examined that currently comprise sections *Pentadenia* and *Stygnanthe* (Fritze & Williams 1988). The majority of species examined were from sect. *Stygnanthe*; the exceptions are *C. strigosa, C. nervosa,* and a morphological variant of *C. strigosa*, named *C. kucyniakii*, which was treated as a species of *Columnea* sensu Wiehler. Fritze and

Williams (1988) comment that the pollen of *C. kucyniakii* did not fit within the variation of sect. *Columnea* but was more similar to that of sect. *Collandra* (*Dalbergaria* sensu Wiehler) or sect. *Ortholoma* (*Trichantha* sensu Wiehler). The table summarizing the pollen characters shows that the pollen of *C. kucyniakii* does not appear to be the same as that of *C. strigosa*. This is most likely explained as intraspecific variation. Other species in which more than one individual was used exhibited a fair amount of intraspecific variation, indicating that pollen morphology among these species will be useful only when numerous individuals are examined.

Gynoecium. Ovaries in the species of *Columnea* included here are conical and may be glabrous, pilose, tomentose, sericeous, or villous. The hairs are uniseriate and generally transparent but red in a few species. Styles are white, pale pink, red, or yellow, and range from glabrous to slightly pilose, frequently becoming capitate-glandular distally. The stigma is either bilobed or stomatomorphic, and white, green, red, or yellow. Bilobed stigmas are usually papillose.

Nectary. The nectary structure in *Columnea* has been considered taxonomically useful (Brown 1938; Wiehler 1983; Kvist & Skog 1993). Wiehler (1973, 1983) considered the presence of a five-lobed nectary, occasionally with the two dorsal lobes enlarged or fused, of such importance that he segregated the species of sections *Pentadenia* and *Stygnanthe* into a separate genus, *Pentadenia*. Kvist and Skog (1993) placed less emphasis on the nectary and assigned some species that were morphologically similar to the species in Wiehler's genus *Pentadenia* in sect. *Stygnanthe*. Wilson (1974) described the floral anatomy of *Columnea* accession G361 from the L. H. Bailey conservatory at Cornell University (BH). This plant, vouchered as *H. E. Moore 7773*, has been identified as *C. angustata*. Some flowers from this individual had the typical five-lobed nectary structure, but a series of others demonstrated a continuum of reduction of the three separate glands and enlargement and fusion of the two dorsal glands with a corresponding reduction and enlargement of vascular tissue to these glands (Wilson 1974).

Examination of several field-collected populations of *C. strigosa* for this study, and by Kvist and Skog (1993), showed a wide range in variation from the five-lobed structure to the presence of only a dorsal unlobed gland. Examination of several flowers from single individuals of *C. orientandina*, *C. angustata,* and *C. colombiana* grown in the greenhouses at the University of Wisconsin–Madison also showed variability in the nectary structure. *Columnea orientandina* exhibited the widest range in variation corresponding to the variability seen in populations of *C. strigosa*.

Although it may be that only species of sections *Pentadenia* and *Stygnanthe* have the potential for the five-lobed nectary structure, the extreme variability of this trait, especially within individuals, drastically limits its taxonomic utility. This character is not recommended for taxonomic delimitation at any level in *Columnea*.

Fruit. The fruit of *Columnea* distinguishes it from two morphologically similar genera, *Alloplectus* and *Drymonia*. *Columnea* fruits are indehiscent berries, either white or colored, generally pale pink to lavender, but darker purple in some species. Fruits of *Alloplectus* and *Drymonia* are fleshy but dehiscent. The valves of the fleshy capsules of *Alloplectus* and *Drymonia* open to reveal a generally brightly colored mass of seeds and funiculi. Because the interior surfaces of the valves are also frequently brightly colored, the entire fruit is highly conspicuous, meriting the name of "display fruit" (Wiehler 1983).

In the species of *Columnea* located basally in the cladograms derived from morphology, cpDNA restriction site variation, and combined analysis of these data (Fig. 2) (Smith & Sytsma 1994 a, b, c), the fruit appears to be intermediate between the capsular display fruit of *Alloplectus* and the indehiscent berry of *Columnea*. *Alloplectus* is considered the sister genus to *Columnea*, and the fruit type in these basally located species is considered to be transitional. The transitional fruit is found in *C. trollii* and *C. oblongifolia*. Fruits collected in the wild were indehiscent, and when pressure was applied to the fruits by squeezing, the fruit became detached from the receptacle but did not split. Upon drying, however, these fruits split open evenly along the loculicidal lines. Although the fruits do not dehisce in the wild, they have retained the ability to dehisce, but only do so when dried.

Except for the interesting fruits of *C. trollii* and *C. oblongifolia*, the berries of sections *Pentadenia* and *Stygnanthe* are indehiscent and either globose or ovoid, white, pale lavender, or pink. The fruits of *C. strigosa*, *C. ultraviolacea,* and *C. lavandulacea* are a darker purple. Berry shape and color is constant within most species but is of no apparent taxonomic utility above the species level. The berries of *C. byrsina* are the only fruits that are variable within a species. Collections made in the Napo province of Ecuador have ovoid berries, whereas all other collections exhibited globose fruits.

Seeds. In a survey of seed morphology of over 700 species of Gesneriaceae, Beaufort-Murphy (1983) concluded that the seeds of *Columnea* formed an "unusually uniform group." The seed shapes, apical characters, cell orientation, and cell shape, as well as primary and secondary cell characters, were essentially the same for all species examined, which included representatives from five of the six currently recognized sections.

Seeds of *Columnea* are narrowly elliptic, with slightly spiral to spiral cell orientation and with acute apices. The primary and secondary cell characters are: edges elevated and basically smooth to rough, faces depressed and smooth to rough, and crests discrete (Beaufort-Murphy 1983).

HABITATS AND DISTRIBUTION

All species of *Columnea* are tropical; the majority are found in montane to cloud forest habitats but many species occur in the lowland rain forests as well. The species of sections *Pentadenia* and *Stygnanthe* occupy a wide range of habitats. In general, sect. *Pentadenia* is found at higher elevations than sect. *Stygnanthe*. The species that occur at the highest elevations in the entire genus are in sect. *Pentadenia*; *C. strigosa, C. trollii, C. hypocyrtantha,* and *C. oblongifolia* all occur in habitats above 2500 m. In contrast, it is rare to find species of sect. *Stygnanthe*, or any other sections, above this elevation. The higher-elevation species are found frequently in dwarf woodlands and cloud forests in association with *Weinmannia* (Cunoniaceae) and *Clusia* (Clusiaceae). Species at lower elevations occur in montane forests to cloud forests or similar habitats that are very wet.

The exception to this elevational distribution is a group of species in sect. *Pentadenia* that grows in seasonally dry forests of Ecuador, Mexico, and Panama (Harling 1979; Rzedowski 1981). These species are all characterized by clusters of leaves at the stem apices and an alternation of series of long and short internodes along the stem. Apparently the internodes grow longer during the wet season. During the dry season, the plants flower and drop all but the leaves at the apex; minimal apical growth creates series of short internodes (Mansfeld 1937; Kvist & Skog 1993).

Although no species of *Columnea* is particularly well adapted for disturbance, some species appear to be more tolerant of moderate disturbance than others. Species in sect. *Pentadenia* appear to be intolerant of disturbance. These species are becoming less frequently collected, at localities mostly in more pristine habitats. Several collection sites for *C. strigosa* were revisited during my field work. In cases where plants had been collected 3–5 years previously, and the sites had been highly disturbed, no individuals of *C. strigosa* were found, suggesting that the populations had since become extinct locally.

Tolerance for disturbance appears to be much greater in sect. *Stygnanthe*. Several collections of *C. spathulata, C. angustata,* and *C. byrsina* have been found with cultivated species such as citrus or cacao. During field work for molecular analysis, several of these species were easily obtained by collecting along roadsides with standing trees or in pastures. *Columnea inconspicua* and *C. rileyi* were both collected growing together in a roadside *Agave* hedge in Ecuador. Many of the earlier herbarium labels for these species identify roadsides or pastures as collecting sites. Despite the occurrence of these species in disturbed habitats, the viability of their populations in these conditions is almost entirely unknown.

With the exception of *C. strigosa*, *C. angustata, C. spathulata, C. trollii,* and *C. nervosa*, fewer than 10 collections for each species were available for this study. The collections for *C. nervosa* were not numerous from any specific locality, but was compensated for by this species' widespread distribution (Fig. 10). Most of the species included in this study are local endemics; the exceptions are the species listed above, excluding *C. trollii*. To some extent the rarity of these species may be due to lack of collections. Further exploration is likely to discover more populations of these species, but probably not in sufficient numbers to consider them common.

Section *Pentadenia* is represented from southern Mexico southward to Bolivia, whereas sect. *Stygnanthe* is found from Costa Rica southward to Bolivia, eastward into Venezuela, and in Puerto Rico (Fig. 1). The majority of species are found on west-facing slopes of the Andes. *Columnea orientandina*, *C. poortmannii,* and populations of *C. angustata, C. byrsina, C. inconspicua, C. rileyi*, and *C. lavandulacea* are the exceptions and occur on the eastern side of the Andes as well (Figs. 5, 8, 17–19).

As with other Andean taxa (Vuilleumier 1969; Berry 1982; Molau 1988; Grifo 1989; Luteyn 1989), the species composition of *Columnea* is different on either side of the Huancabamba deflection. For example, *C. strigosa* is a common, widespread species found throughout the northern Andes from Venezuela to northern Peru, but has never been collected south of the Piura divide. Only two species, *C. orientandina* and *C. spathulata*, (Figs. 8, 24) cross the barrier, and neither of these species is strictly an Andean taxon but is frequently found in lowland habitats.

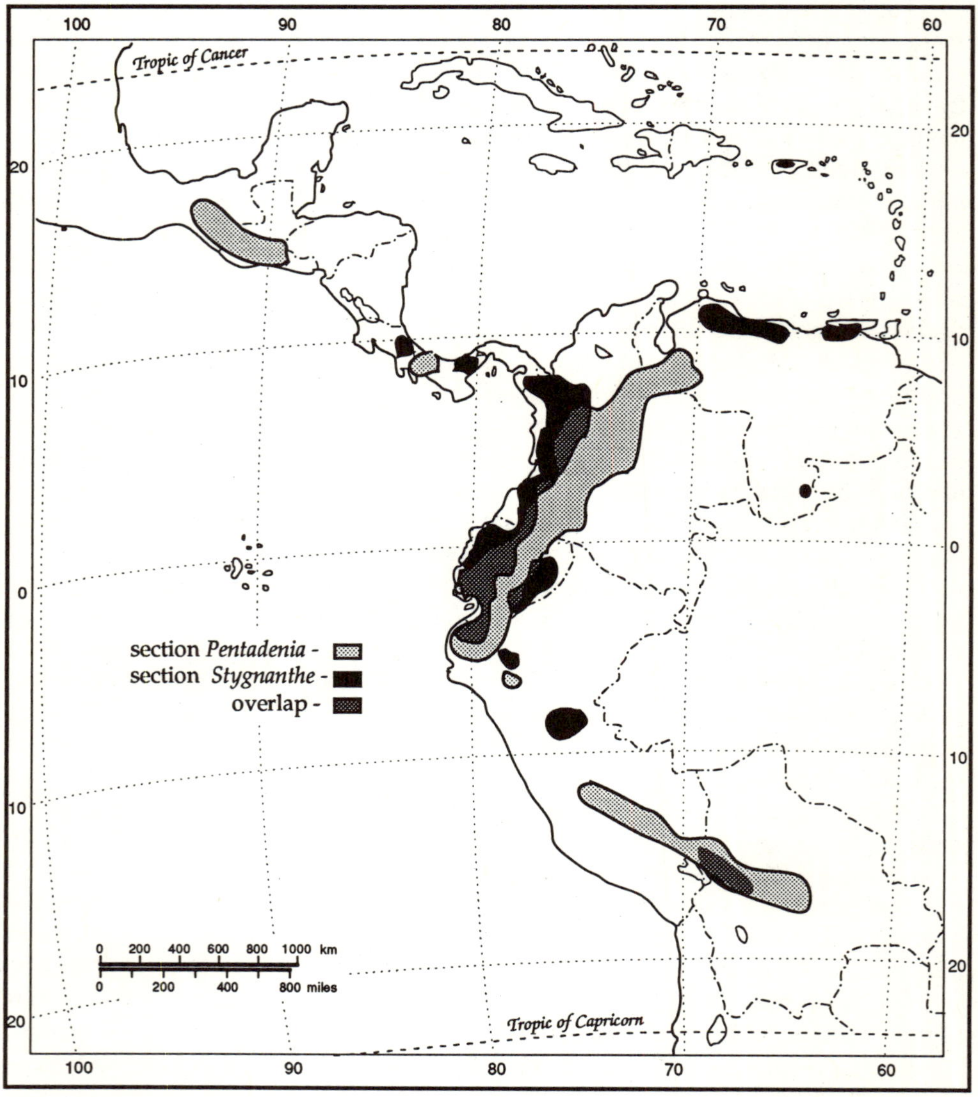

FIG. 1. Distribution of *Columnea* sections *Pentadenia* and *Stygnanthe* in Central and South America.

Several Andean taxa are readily divided into northern, central, and southern Andean groups (Berry 1982; Molau 1988; Grifo 1989). This division applies to most species of *Columnea* sections *Pentadenia* and *Stygnanthe*, with the exception of the widespread species *C. spathulata*.

Several geographic disjunctions occur within species of *Columnea* that are most likely the result of long-distance dispersal by birds. *Columnea nervosa* is found only in southern Mexico and neighboring Guatemala and El Salvador, Panama, and southern Ecuador. There are subtle morphological variants corresponding to geographic regions, but they are insufficient to merit species status. For example, the collections from Ecuador have red corollas with a yellow apex. These corollas are also larger than those of the Central American representatives, ranging from 3.3–6.0 cm long, 0.8–1.4 cm wide at the widest point. The corollas of populations from Central America differ slightly from the Ecuadorian collections, but fit within the range of variation seen in Ecuador. Corollas of the Central American collections are smaller, 2.7–4.0 cm long, 0.6–1.0 cm wide; those of the Panamanian collections all have yellow corollas. Much of the morphological variation seen in the Central American populations is a subset of the variation of the Ecuadorian populations. Therefore, it is hypothesized that the Central American representatives are derived, via a long-distance dispersal event, from the Ecuadorian populations. This movement northward is also supported by phylogenetic evidence (Fig. 2). Most of the basally located lineages of sect. *Pentadenia* have a southern Peruvian or Bolivian distribution (Smith & Sytsma 1994a, b, c). Other species in sect. *Pentadenia* are from more recently derived lineages within the genus and are from Ecuador. A movement from Ecuador to Central America is an extension of the northerly movement within the section and genus as a whole. Therefore, the dispersal of *C. nervosa* from Ecuador to Central America is more plausible than the reverse based on the subset of morphological variation found within Central America and the evidence from the phylogenetic analyses.

Early South America to North America migrations are known for several other groups, such as Cactaceae, Loasaceae, Nyctaginaceae, and Zygophyllaceae (Raven & Axelrod 1974). *Fuchsia* (Onagraceae) is a southern group, and sect. *Skinnera*, the sister group to all other species in the genus, occurs in New Zealand and the Pacific (Berry 1982; Sytsma & Smith 1988; Sytsma et al. 1991). However, one of the American sections of *Fuchsia* to branch early in the cladogram is sect. *Jimenezia* (Berry 1982; Smith & Sytsma 1988), which is found in Central America. This group was presumably dispersed there early in the evolution and radiation of the genus (Berry 1982). The appearance of *C. nervosa* in Mexico and Central America parallels the distribution pattern seen among sections of *Fuchsia*.

The most widespread species of either sect. *Pentadenia* or sect. *Stygnanthe* is *Columnea spathulata*. Three species have been proposed for this highly variable taxon: *C. microsepala*, representing the Venezuelan populations, *C. zapotalana*, encompassing the southern Ecuadorian populations, and *C. spathulata,* comprising the northern Ecuadorian and Colombian populations. In general, the populations from northern Ecuador and Colombia are in somewhat wetter habitats, and the morphology of these populations is distinctive. However, examination of all populations throughout the range demonstrate a continuum of variation, linking the populations and morphological variants into a single species.

Columnea spathulata s.l. is common in Ecuador and the coastal range of Venezuela, but is rarely collected elsewhere. If the current understanding of the distribution of *C. spathulata* is correct, and not an artifact of insufficient collecting, the few populations lying outside of the major distribution centers may represent recent long-distance dispersal events. The center of origin for *C. spathulata* is likely to be in Ecuador. A dispersal event from Ecuador to the coastal range of Venezuela could have resulted in the populations there. More recent, occasional dispersal events may explain populations in Bolivia, Colombia, and southern Venezuela (Fig. 24). Unfortunately, these key areas are also poorly collected regions, and only further botanical exploration will resolve whether these outlying localities harbor reproducing populations or merely dispersed individuals.

BREEDING SYSTEMS

Observations were made regarding the breeding system in *Columnea* on individuals grown in the greenhouse at the University of Wisconsin–Madison. Species of *Columnea* examined in this study (see appendix) are protandrous, the anthers dehiscing before the corolla is fully open. Anthers are presented for two to three days and are then retracted into the tube of the corolla by the curling of the filaments. During anther retraction the style elongates, and the stigma becomes receptive and occupies the former position of the anthers. A single individual of *C. angustata* from Panama, accession number W2434, deviated from this behavior. In this individual, protandry had been lost, and the anthers and stigma were receptive simultaneously before the corolla was fully open. In the greenhouse, this individual frequently set fruit without hand-pollination. These self-pollinations occurred during the winter when potential pollinators were not present in the greenhouse.

Several attempts were made to self-pollinate the accessions of *Columnea* in cultivation at the University of Wisconsin–Madison (appendix). Although many pollinations resulted in aborted fruits, every accession that produced mature fruits had viable seed (Smith 1991). The selfed fruits of *C. colombiana* produced over 200 seeds per fruit, and resulted in nearly 100% germination (Smith 1991). A similar rate of germination was seen with *C. inconspicua* (Smith 1991). Self-fertilization has been observed in *C. byrsina* (Ertelt 1990) and in *C. ambigua*, which vigorously fruits in the greenhouses at the Montréal Botanical Garden (pers. obs.).

HYBRIDS

Several interspecific hybrids in sect. *Stygnanthe* were produced during this study, but only one flowered (Smith 1991). In general, hybrids were much easier to induce from species that were closely related, i.e., within the same phylogenetic clade, than those that were not (Smith 1991). Intersectional crosses between sections *Stygnanthe* and *Columnea* initially produced fruits, but these always aborted before maturity (Smith 1991).

Although it has been thoroughly demonstrated that *Columnea* hybridizes in the wild (Morley 1971, 1976b; Byrne & Morley 1976), no apparent hybrids were detected in sections *Pentadenia* or *Stygnanthe* through examination of herbarium material, live plants in the field, or molecular techniques (Smith & Sytsma 1994a, b, c).

PHYLOGENY

An integral aspect to the study of *Columnea* sections *Pentadenia* and *Stygnanthe* was a cladistic analysis of data from morphology and chloroplast DNA (cpDNA) restriction sites. The detailed methods, choice of characters, enzymes used, and analysis are presented elsewhere (Smith & Sytsma 1994a, b, c), but a summary of the results is presented here.

Three separate cladistic analyses were performed on the species in these two sections. Species of *Alloplectus* and *Drymonia* were used as outgroup taxa in all analyses. The first analysis was based on morphological characters. Data were collected from all parts of the plant for a total of 35 characters. These data were analyzed in PAUP version 3.1 (Swofford 1991) using Wagner parsimony, which produced 10 most-parsimonious trees (Smith & Sytsma 1994a). The strict consensus of these 10 most-parsimonious trees revealed several trends in the evolution of *Columnea*. For example, the basal-most lineages in the phylogeny were species from southern Peru and Bolivia, which suggests a central-southern Andean origin for the genus as a whole. Morphological variants of *C. strigosa* and *C. spathulata* were included in this analysis to determine the phylogenetic relationships of these taxa. For both species, all variants appeared as a monophyletic group in the strict consensus tree, thereby providing further justification for combining several of the segregates in single species.

The second analysis used cpDNA restriction site variation as a data source. The DNA from *Columnea* species was digested with 41 different restriction enzymes to produce 296 characters. These data were also analyzed in PAUP 3.1 (Swofford 1991) using the same methods as were used with the morphological data. This analysis resulted in 4045 most-parsimonious trees (Smith & Sytsma 1994b). (For further explanation of the large number of trees, see Smith & Sytsma 1994b). The strict consensus of these 4045 most-parsimonious trees revealed many of the same trends found with the morphological data. The basal-most lineages of the tree also represented species from southern Peru and Bolivia, which again suggested a central-southern Andean origin for the genus. Numerous individuals from distinct populations were used for some species. All of these were revealed as belonging to monophyletic groups, including the different morphological variants of *C. strigosa* and *C. spathulata*.

Various methods were used to compare the topologies of the two analyses. All of these indicated that the two topologies were congruent, and therefore a third analysis was performed with all of the data combined (Smith & Sytsma 1994c). For this third analysis PAUP 3.1 (Swofford 1991) also used and produced 28 most-parsimonious trees. The strict consensus of these 28 most-parsimonious trees is presented in Fig. 2. The combined analysis did not use all species of sections *Pentadenia* and *Stygnanthe*, because some species were not obtained for DNA analysis. However, this strict-concensus tree represents the most complete analysis in terms of available data. The topology of the combined-analysis tree is similar to both the cpDNA and morphology trees and contains aspects of both trees. No single data set had an overwhelming effect on the combined analysis. Morphological data provided support for the phylogeny in some areas, and the cpDNA provided support for others (Smith & Sytsma 1994c).

SPECIES AND SPECIATION

Nearly every species in *Columnea* is distinguished by a suite of characters derived from both reproductive and vegetative organs. The pattern of distribution of individual character states throughout the family necessitates the use of suites of characters in defining species and groupings at higher taxonomic levels. Therefore, in defining species of *Columnea* it was impossible to use any single character as a defining feature of a species or group of species. Every species was defined on the basis a group of characters. This polythetic species concept has resulted in homoplasy for the morphological data; however, any other method of defining

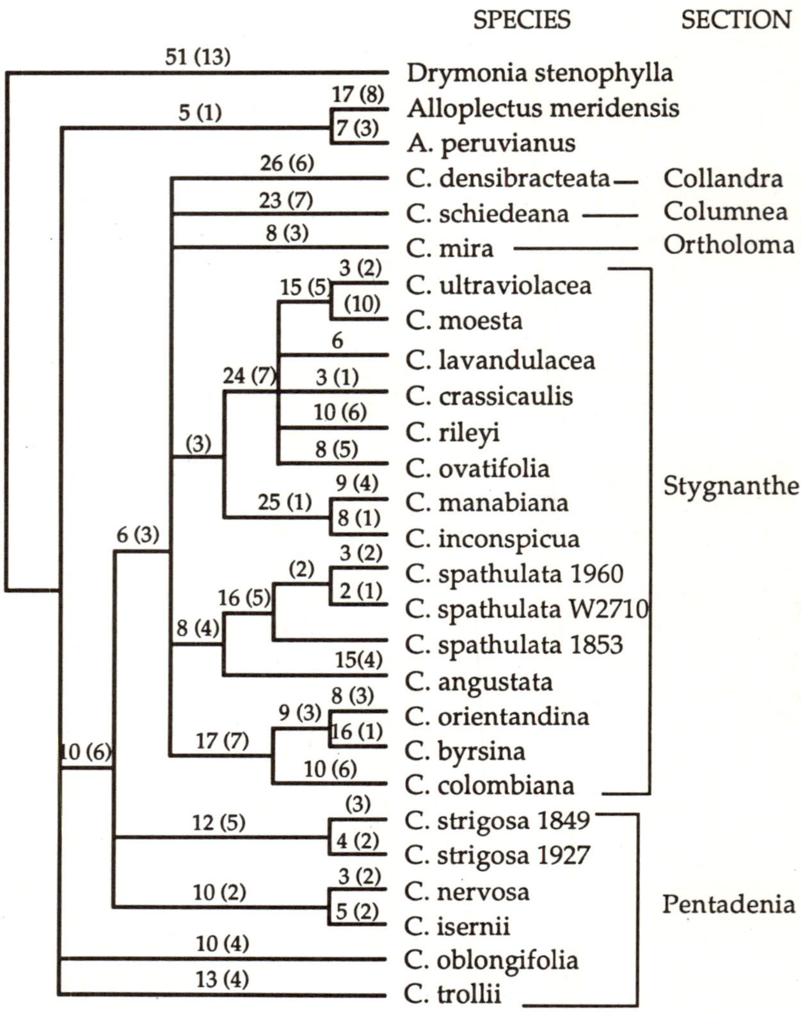

FIG. 2. Strict consensus of 28 most-parsimonious trees from a cladistic analysis using a combined data set of morphological and cpDNA restriction site variation. Numbers along clades indicate the number of character state changes that support that clade. Numbers in parentheses are the character state changes that are homoplastic. For further information regarding characters and methods of analysis, see Smith and Sytsma, 1994c.

species will not accurately circumscribe the full morphological variation that can be found within species or groups of species.

Because there is a wide range of morphological variation found within many species included in this study, delimitations of species relied on clear breaks in morphology for all the characters used to define the species. For example, the combination of *C. microsepala* and *C. spathulata* into a single species was necessary on the basis of the pattern of morphological variation found among the populations of these taxa. There are numerous populations of these two morphological variants that are readily assigned to one or the other segregate species; however, there are an equal number of populations that are intermediate between the two for one or more characters. Because there is a continuum of variation among the variants, it is necessary to recognize them as a single species.

This pattern of diversity applies not only to species but higher levels as well. For example, anisophylly is one of the primary characters used to distinguish sect. *Collandra* from other sections. Yet, anisophylly occurs in sections *Columnea, Stygnanthe,* and *Ortholoma*, as well as in other genera in the Gesneriaceae, such as *Monopyle* Moritz ex Benth., *Paradrymonia* Hanst., and *Streptocarpus* Lindl.

The most likely explanation for the homoplasy seen in *Columnea* is a recent origin for the species within the genus as indicated by several characters. Characters, such as chromosome number (Morley 1972), seed morphology (Beaufort-Murphy 1983), chemical constituents (Harborne 1967; Kvist & Pedersen 1986), and nodal anatomy (Wiehler 1983), vary only slightly, if at all, below the tribal level. Other characters that have slightly more variation, such as pollen morphology (Fritze & Williams 1988) and cpDNA variation (Smith & Sytsma 1994b), are not as variable as they are in other families. All of these factors indicate either a recently derived group or one that for some reason (i.e., selection) is resistant to differentiation. Selection on such a diverse group of characters to prevent differentiation seems unlikely. Also, species of sections *Pentadenia* and *Stygnanthe* are primarily distributed in the Andes, a mountain range whose origin is hypothesized at ca. 6 mya (Hammen 1974; Simpson 1975, 1979). Although some of the species inhabit lowlands, the most basally located species of the phylogeny are high-elevation Andean species (Smith & Sytsma 1994a, b), which suggests that the genus itself is not older than the Andean range.

A recently derived group may not have had sufficient time for mutation of genes and alleles controlling various traits to accumulate. As a result, when populations became isolated during Andean orogenies, various genes and alleles may have become fixed or lost in different combinations in different populations. Such a scenario could explain the pattern of morphological variation seen in species of *Columnea* today.

The high morphological diversity of *Columnea* parallels the pattern of diversity seen for *Fuchsia* sect. *Fuchsia* (Berry 1982). Like *Fuchsia*, *Columnea* had an austral South American origin, as suggested by cladistic analyses (Smith & Sytsma 1994a, b, c). These early habitats would have been reasonably cool and perhaps caused selection for adaptations for survival at higher elevations. Climatic changes during the Andean orogenies would have created new, open habitats suitable for exploitation and rapid speciation (Raven 1980; Berry 1982).

TAXONOMY

Columnea L., Sp. pl. 638. 1753.—TYPE: *Columnea scandens* L.

Achimenes P. Browne, Civ. nat. hist. Jamaica 270, pl. 30. 1756, non *Achimenes* Persoon, Syn. pl. 2: 164. 1806.—TYPE: *Achimenes major, herbacea subhirsuta, oblique assurgens* P. Browne [=*Columnea fawcettii* (Urban) C. Morton].

Dalbergaria Tussac, Fl. Antill. 1: 141, pl. 19. 1808–13.—TYPE: *Dalbergaria phoenicea* Tussac [=*Columnea sanguinea* (Persoon) Hanstein].

Vireya Rafinesque, Specchio sci. 1: 194. 1814.—TYPE: *Vireya sanguinolenta* Rafinesque [=*Columnea sanguinea* (Persoon) Hanstein].

Eusynetra Rafinesque, Fl. tellur. 2: 57. 1837.—TYPE: *Eusynetra bicolor* Rafinesque [=*Columnea hirsuta* Swartz].

Glycanthes Rafinesque, Sylva tellur. 83. 1838.—TYPE: *Glycanthes scandens* (L.) Rafinesque [=*Columnea scandens* L.].

Trichantha Hooker, Icon. pl. 3: pl. 666. 1844.—TYPE: *Trichantha minor* Hooker [=*Columnea minor* (Hooker) Hanstein].

Collandra Lemaire, Fl. Serres Jard. Eur. 3: pl. 223. 1847.—TYPE: *Collandra pilosa* Lemaire [=*Columnea aureonitens* Hooker].

Ortholoma (Bentham) Hanstein, Linnaea 26: 184, 209. 1854.—TYPE: *Ortholoma acuminatum* (Bentham) Hanstein [=*Columnea acuminata* Bentham].

Pentadenia (Planchon) Hanstein, Linnaea 26: 187, 211. 1854.—TYPE: *Pentadenia aurantiaca* (Decaisne ex Planchon) Hanstein [=*Columnea strigosa* (Benth.) Hanst.].

Pterygoloma Hanstein, Linnaea 26: 188, 211. 1854.—TYPE: *Pterygoloma repens* (Hooker) Hanstein [=*Columnea repens* (Hooker) Hanstein].

Stenanthus Oersted ex Hanstein, Linnaea 26: 184, 209. 1854.—TYPE: *Stenanthus heterophyllus* Oersted ex Hanstein [=*Columnea grata* C. Morton].

Stygnanthe Hanstein, Linnaea 26: 185, 209. 1854.—TYPE: *Stygnanthe moesta* (Poeppig) Hanstein [=*Columnea moesta* Poeppig].

Fluckigeria Rusby, Bull. Torrey Bot. Club 21: 488. 1894, non *Flueckigeria* Kuntze, 1891. *Kohlerianthus* Fritsch in Engler & Prantl, Nat. Pflanzenfam. Nachtr. 1: 300. 1897, nom. nov.—TYPE: *Fluckigeria fritschii* Rusby [=*Columnea fritschii* (Rusby) J. F. Smith].

×*Colbergaria* Wiehler, Selbyana 1: 408. 1976.

×*Coltrichantha* Wiehler, Selbyana 1: 408. 1976.

×*Coltadenia* Wiehler, Selbyana 1: 409. 1976.

×*Daltrichantha* Wiehler, Selbyana 1: 409. 1976.

×*Trichadenia* Wiehler, Selbyana 1: 409. 1976, non *Trichadenia* Thwaites, 1855. ×*Tricanthenia* Wiehler, Selbyana 6: 116. 1983, nom. nov.

Bucinella Wiehler, Selbyana 2: 91. 1977, non *Bucinella* A. Fucini, 1936. *Bucinellina* Wiehler, Selbyana 5: 381. 1981, nom. nov.—TYPE: *Bucinella nariniana* Wiehler [=*Columnea nariniana* (Wiehler) L. P. Kvist & L. E. Skog].

Herbs or herbaceous shrubs with erect, ascending, pendent, creeping, or scandent stems, epiphytic or terrestrial. Leaves opposite and isophyllous to strongly anisophyllous, or whorled, petiolate or sessile, apex acute to acuminate, base oblique, adaxial surface generally green, sometimes purple, abaxial surface green, mottled with red-purple, with red-purple apices, or entirely purple, adaxially and abaxially glabrous to densely sericeous to lanate, margins entire to serrate or crenate. Inflorescences axillary, of 1–12 flowers, bracteate; bracts large or inconspicuous, frequently caducous. Pedicel 0.01 to 29.8 cm, erect or pendent, occasionally with oval-round dark purple glandular structures near the calyx. Calyx generally clasping corolla, sometimes with the reflexed tips of the lobes; lobes generally equal to subequal, linear to ovate, glabrate to sericeous or lanate, generally with fewer hairs or glabrous on adaxial surface of the lobes, margins entire to dissected, color and margin frequently matching that of the leaves. Corolla tubular, gibbous at base, expanding in the middle and constricted near the opening in sections *Pentadenia* and *Stygnanthe*, expanding in other sections; lobes five, generally free, occasionally the two dorsal lobes fused into a galea, generally semiorbicular and equal to subequal in sections *Pentadenia* and *Stygnanthe*, semiorbicular to lanceolate in other sections, indumentum sparse to villous or sericeous on exterior surface, glabrous to hirsute or capitate-glandular on interior, generally glandular dorsally and near opening. Stamens 4; filaments included or rarely exserted, connate at base and adnate to corolla tube at base; anthers quadrate to rectangular, coherent at apices, dehiscing by longitudinal slits, glabrous to sparsely hirsute or capitate-glandular. Ovary conical, glabrous to sericeous; style white, pale pink, red, or yellow, glabrous to sericeous, frequently with glandular trichomes, stigma bilobed or stomatomorphic, included in corolla, rarely exserted. Nectary varying from five free equal or subequal glands to a single dorsal gland resulting from the fusion of two glands. Fruit a globose or ovoid berry, glabrous to sericeous, indehiscent or rarely splitting open upon drying; seeds narrowly elliptic to elliptic, striate, ca. 1 mm long. Chromosome number: n = 9.

Key to the Sections of Columnea

1. Mature fruit a flattened white berry; corollas small. *Columnea* sect. *Bucinellina*.
1. Mature fruit a globose to ovoid white or colored berry; corollas large or small.
    2. Leaves very long, generally more than 16 cm, strongly anisophyllous. *Columnea* sect. *Collandra*.
    2. Leaves less than 16 cm long, isophyllous to strongly anisophyllous.
        3. Corollas widening from the base and bilabiate, never ventricose or constricted at the mouth. *Columnea* sect. *Columnea*.
        3. Corollas ventricose, not bilabiate.
            4. Corollas with appendages between lobes, or with indumentum completely obscuring the corolla tube. *Columnea* sect. *Ortholoma*.
            4. Corollas without appendages between lobes, the corolla tube visible beneath indumentum.
                5. Corollas slightly ventricose; pedicels mostly less than 2 cm long; floral bracts less than 2 cm long. *Columnea* sect. *Stygnanthe*.
                5. Corollas strongly ventricose; pedicels more than 2 cm long, or if less, then floral bracts more than 2 cm long. *Columnea* sect. *Pentadenia*.

Key to the Species of Columnea Section Pentadenia and Section Stygnanthe

Note: Leaf measurements for this study were made using the widest part of the leaf for the width; the length was measured as the length of the midrib from petiole to apex. Petioles were measured separately. Figure 3 shows the positions at which measurements of the calyces and corollas were taken.

1. Leaves in whorls of four.
    2. Corollas bright red-purple, densely long hirsute, 10–12 mm wide at widest point; laminas and calyx lobes green. 16. *C. fritschii*.
    2. Corollas yellow, densely sericeous, 5.5–7 mm wide at widest point; undersides of laminas and calyx lobes purple. 26. *C. ultraviolacea*.
1. Leaves opposite or appearing alternate, not whorled.
    3. Most pairs of leaves strongly anisophyllous, the larger leaf of a pair at least twice as long as the smaller.
        4. Leaf margin serrate or crenate to crenulate.
            5. Margin of calyx lobes entire, calyx lobes spatulate to lanceolate; adaxial leaf surface green but purple at apex or entirely purple, leaf margin crenate. 24. *C. spathulata*.

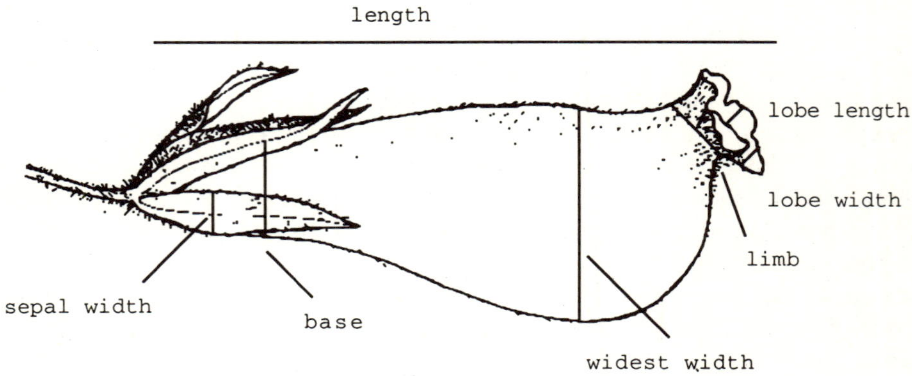

FIG. 3. Diagram of corolla of *Columnea* showing flower measurements used in this study.

5. Margin of calyx lobes serrate or coarsely toothed, calyx lobes spatulate or elliptic to lanceolate; adaxial leaf surface green, leaf margin crenate to serrate.
   6. Larger laminas mostly more than 5 cm long (3.5–11.5 cm), oblong to elliptic; stems sublignose; calyx lobes elliptic to lanceolate; corollas yellow; Puerto Rico. 10. *C. ambigua*.
   6. Larger laminas mostly less than 5 cm long (1.4–5.0 cm), ovate to elliptic, stems more or less woody; calyx lobes spatulate; corollas yellow with orange spots on lobes; Colombia. 25. *C. suffruticosa*.

4. Leaf margin entire.
   7. Corollas bright red with yellow limbs; stamens and stigma exserted 5–9 mm beyond opening of corolla tube. 13. *C. byrsina*.
   7. Corollas yellow, cream, or lavender; stamens and stigma enclosed, if exserted, then only slightly beyond opening of corolla tube (*C. orientandina*).
      8. Corollas lavender (rarely cream) with green limbs and purple spots on lobes; larger laminas ovate to obovate, 1.7–5.5 cm long, lamina length less than twice the width; indumentum of laminas and stems lavender. 18. *C. lavandulacea*.
      8. Corollas yellow; laminas lanceolate to elliptic to oblanceolate, usually more than 6 cm long (3.5–17 cm), lamina length more than twice the width, indumentum of laminas and stems not lavender.
         9. Laminas elliptic to oblanceolate, not falcate, apex generally obtuse, not acuminate.
            10. Leaf margin crenate to entire, laminas pilose to villous abaxially; calyx lobes spatulate to lanceolate; corollas yellow, not waxy, lacking glandular tipped hairs; filaments pilose; berries globose, white. 24. *C. spathulata*.
            10. Leaf margin entire, laminas sericeous to lanate abaxially; calyx lobes lanceolate to subulate; corollas lemon-yellow, waxy, with glandular tipped hairs; filaments glabrous; berries ovoid, pink. 21. *C. orientandina*.
         9. Laminas lanceolate, slightly falcate, the midrib curving at the apex, apex sharply acuminate, never obtuse.
            11. Floral bracts conspicuous, ovate, 13–19 mm long, 5–9 mm wide; ovaries and styles pubescent to pilose. 19. *C. manabiana*.
            11. Floral bracts lanceolate, inconspicuous or caducous, up to 5 mm long, 1.2 mm wide; ovaries and styles glabrous. 17. *C. inconspicua*.

3. Leaves isophyllous, or if slightly anisophyllous, the larger leaf not twice as long as the smaller.
   12. Corollas bilabiate, lobes unequal, more than 4 mm long, some lobes reflexed, the two dorsal lobes fused.

13. Corollas bright orange, rarely red-orange with yellow apex, lateral lobes slightly and ventral lobe strongly reflexed; calyx lobes ovate to lanceolate, the margin serrate; pedicels 1.7–30 cm long; northern Peru to Venezuela, 1500–3800 m. 8. *C. strigosa*.

13. Corollas yellow, occasionally red-purple with a yellow limb, lateral lobes reflexed, ventral lobe not reflexed; calyx lobes linear to lanceolate, the margin entire; pedicels 2.9–9 cm long; southern Peru and Bolivia, 2000–2900 m. 6. *C. oblongifolia*.

12. Corollas not bilabiate, lobes equal to subequal, mostly less than 4 mm long, none reflexed or fused.

14. Laminas lanceolate or oblanceolate, longest laminas more than 4 times as long as wide, acuminate.

15. Margin of calyx lobes serrate to serrulate. 11. *C. angustata*.

15. Margin of calyx lobes entire.

16. Corollas crimson with purple spots on lobes, villous to eglandular-sericeous. 27. *C. xiphoidea*.

16. Corollas pale yellow, without spots, pilose and toward limb glandular-sericeous. 17. *C. inconspicua*

14. Laminas ovate, orbicular, or elliptic, if somewhat lanceolate then not longer than 4 times the width, acute to acuminate.

17. Expanded portion of corolla at least twice as wide as constriction near opening.

18. Margin of calyx lobes serrate to dentate.

19. Laminas ovate, up to 6 cm long; ovaries glabrous; filaments pubescent. 7. *C. poortmannii*.

19. Laminas ovate to elliptic, up to 18 cm long; ovaries sericeous; filaments glabrous. 5. *C. nervosa*.

18. Margin of calyx lobes entire to minutely denticulate.

20. Calyx lobes ovate; pouch of corolla extending forward beyond the opening of the corolla; ovaries and styles glabrous. 2. *C. hypocyrtantha*.

20. Calyx lobes lanceolate to linear; pouch of corolla centered to slightly forward but not extending beyond the opening of the corolla; ovaries pilose to lanate; styles glabrous, sparsely pilose to hirsute, or glandular.

21. Calyx lobes linear, 1–3.5 mm wide; corollas red-violet, the limb green and the lobes with purple spots; pedicels 1–1.5 (–2.8) cm long; laminas elliptic, appressed-pilose. 20. *C. moesta*.

21. Calyx lobes lanceolate, 2–8 mm wide; corollas red-orange, the limb yellow; pedicels (0.6–) 2–5.8 cm long; laminas elliptic, rarely lanceolate or ovate, appressed-hirsute, rarely strigose. 9. *C. trollii*.

17. Corolla not swollen, expanded portion of corollas narrow, not twice as wide as constriction near opening.

22. Calyx lobes dissected at base.

23. Laminas ovate to orbicular, the margin entire; stems pendent. 14. *C. colombiana*.

23. Laminas elliptic to lanceolate, the margin serrulate, denticulate, or crenate; stems erect.

24. Pedicels with dark oval to round glands near calyx; corollas red or yellow, the constriction 5–8 mm wide. 5. *C. nervosa*.

24. Pedicels without glands; corollas green-yellow and purple-striped, the constriction 4–6 mm wide. 3. *C. isernii*.

22. Calyx lobes serrate to entire, but not dissected.

25. Margin of calyx lobes serrate to serrulate.

26. Floral bracts conspicuous, 8–32 mm long, sometimes obscuring the inflorescence.

27. Inflorescence subtended by 1–2 floral bracts, these appressed-pilose with golden-yellow trichomes. 1. *C. atahualpae.*
27. Inflorescence subtended by 3 floral bracts, these densely sericeous to villous. 4. *C. lophophora.*
26. Floral bracts inconspicuous or caducous, less than 8 mm long, not obscuring the inflorescence.
28. Corollas 6 mm or more wide, constriction at opening 4 mm or more wide.
29. Leaf margin serrate to dentate. 5. *C. nervosa.*
29. Leaf margin entire to crenulate.
30. Corollas strongly swollen, 7–15 mm wide, red-violet, with green tip and purple spots on lobes. 20. *C. moesta.*
30. Corollas at best only slightly swollen, 2.5–4 (–6) mm wide, red, yellow, or orange.
31. Leaf margin entire, laminas elliptic; stems herbaceous; corollas yellow, orange, or red; calyx lobes lanceolate, margin serrate at base only. 11. *C. angustata.*
31. Leaf margin crenulate, laminas elliptic to ovate; stems more or less woody; corollas yellow; calyx lobes spatulate, margin serrate to apex. 25. *C. suffruticosa.*
28. Corollas less than 6 mm wide, constriction at opening less than 4 mm wide.
32. Leaf margin entire, laminas elliptic; stems herbaceous; corollas yellow, orange, or red. 11. *C. angustata.*
32. Leaf margin crenulate, laminas elliptic to ovate; stems more or less woody; corollas yellow. 25. *C. suffruticosa.*
25. Margin of calyx lobes entire.
33. Floral bracts conspicuous, 8–32 mm long, laminas obovate to elliptic.
34. Inflorescence subtended by 1–2 floral bracts, these appressed-pilose with golden-yellow trichomes. 1. *C. atahualpae.*
34. Inflorescence subtended by 3 floral bracts, these densely sericeous to villous. 4. *C. lophophora.*
33. Floral bracts inconspicuous, less than 8 mm long (if larger, then laminas ovate to orbicular).
35. Laminas ovate to lanceolate, or ovate to orbicular.
36. Laminas ovate to lanceolate; corollas 1.4–3 cm long, red or orange-red.
37. Corollas 2.5–3.0 cm long, red, sparsely pilose to slightly sericeous. 12. *C. antiocana.*
37. Corollas 1.4–2.1 cm long, orange-red, villous. 23. *C. rileyi.*
36. Laminas ovate to orbicular; corollas 3.2–5.2 cm long, yellow or fuchsia-pink to violet.
38. Corollas yellow with red-purple trichomes at opening; stems succulent, creeping; laminas 3.5–6.2 cm long. 15. *C. crassicaulis.*
38. Corollas pink to red-violet, lobes with dark purple spots; stems not succulent, creeping or pendent; laminas 1.5–2.5 cm long. 22. *C. ovatifolia.*
35. Laminas elliptic.

39. Corollas 3.3–4.5 cm cm long, red-violet with a green limb and purple spots on lobes, strongly swollen. 20. *C. moesta.*
39. Corollas 1.4–3.0 cm long, red, orange, or yellow, at best only slightly swollen.
    40. Plants covered with dense white indumentum; corollas orange-red, 1.4–2.1 cm long; calyx lobes spatulate to oblanceolate. 23. *C. rileyi.*
    40. Plants glabrous to sericeous but the indumentum never dense; corollas red, 2.5–3.0 cm long; calyx lobes lanceolate. 12. *C. antiocana.*

Columnea section **Pentadenia** (Planchon) Bentham in Bentham & Hooker, Gen. pl. 2: 1010. 1876. *Columnea* subg. *Pentadenia* Planchon, Fl. Serres Jard. Eur. 6: 45, pl. 552. 1854. *Pentadenia* (Planchon) Hanstein, Linnaea 26: 211, 187. 1854.—Type: *Columnea strigosa* (Bentham) Hanstein.

Suffrutescent, sublignose, terrestrial or epiphytic herbs with erect-ascending (rarely pendent or vining) or sometimes scandent stems to 0.15–3 m tall, 3–12 mm in diameter, terete to squarish, tawny-brown, tan, maroon, green, or red-brown, proximally smooth to scaly and glabrous, distally slightly pilose to densely sericeous; internodes 0.07–14.8 cm long, sometimes shorter internodes clustered at apex; nodes slightly swollen; leaf scars raised to slightly raised. Leaves opposite, sometimes clustered at apex, isophyllous to slightly anisophyllous; laminas 2.3–22.5 cm long, 1.2–13.7 cm wide, ovate to lanceolate or obovate, apex acute or acuminate, base cuneate or rounded, oblique to slightly oblique, adaxially green, red-green, or maroon, slightly pilose to sericeous with uniseriate red or transparent trichomes, or strigose with single-celled trichomes, abaxially silvery green, pale green, maroon, white-pink, or green suffused with purple, glabrate to densely sericeous with uniseriate red or transparent trichomes, or strigose with single-celled trichomes, veins abaxially sericeous, the trichomes frequently longer than on the lamina, lateral veins 4–9 per side; margin entire, denticulate, serrate to serrulate, or crenulate; petioles 0.2–3.6 cm long, green, occasionally red, pilose to densely sericeous. Inflorescences of 1–7 flowers in either axil of a leaf pair; ebracteate or the bracts caducous or with 1–3 floral bracts, equal to unequal, 3–32 mm long, 0.5–16 mm wide, lanceolate to ovate, apex acuminate, occasionally slightly constricted at base, green, pink, or red, sparsely to densely pilose, margin entire. Pedicels absent or up to 29.8 cm long, erect in leaf axil (rarely pendent), green or red-purple, slightly pilose to densely sericeous, eglandular or with 1–several glands near calyx, these sometimes obscured by vesture, ca. 0.4–1.5 mm long, oblong, purplish. Calyx clasping corolla fully, loosely, or only at base and more open towards apex of calyx lobes; lobes equal to unequal, 0.5–4.0 cm long, 0.1–1.2 cm wide, linear to ovate, apex long-acuminate, acuminate, acute, or blunt, green, red, or green with red-purple, red, pink, or maroon at tips, exterior surface glabrate to sericeous, interior surface glabrous to pilose, margin entire to serrate, sometimes ciliate, sometimes dissected at base. Corolla 1.8–9.5 cm long, 5–28 mm wide at widest point, 2–18 mm wide before limb, 2.5–12 mm wide at base, usually tubular, gibbous at base, strongly ventricose to saccate, pale yellow, orange, orange-red, rarely violet-purple, blue, or yellow-green with purple stripes, when red the limb frequently yellow-green, exterior glabrate to densely sericeous, pilose, vesture frequently denser toward limb, interior glabrous to sericeous, frequently with capitate-glandular trichomes dorsally and distally; lobes equal, 7.0–20.0 mm long, 1.5–22 mm wide, semiorbicular to acute; occasionally the corolla bilabiate: the upper two lobes fused into a galea and the lower three lobes reflexed. Filaments connate at base for 3–5 mm, adnate to base of corolla tube for 1–3 mm, white, glabrous to slightly pilose, frequently curling after anthesis; anthers 1–5.5 mm long, 0.8–3.5 mm wide, quadrate, subquadrate, or rectangular, included in corolla tube, rarely exserted. Ovary 3–8 mm long, glabrous to lanate; styles white or red, glabrous or glandular or proximally pilose and distally glandular; stigma bilobed or stomatomorphic, white, green, or red, papillose or smooth, included within corolla tube, rarely exserted. Nectary of 5 free (rarely only 2 dorsal) glands, the 2 dorsal ones sometimes enlarged and connate to slightly connate. Berry 0.4–1.9 cm in diameter, globose, white, purple, pale pink, rose-violet, or rarely green with purple stripes, glabrate to sericeous. Seeds 1.0–1.3 mm long, fusiform, twisted, red-brown, striate.

Wiehler (1973, 1983) segregated the genus *Pentadenia*, based on the five-lobed nectary structure, from *Columnea.* Subsequent studies have demonstrated that this character is variable within an individual (Wilson 1974) and therefore should be less emphasized as a taxonomic character. Kvist and Skog (1993) placed only *C. strigosa* in their sect. *Pentadenia*, and moved all other species of Wiehler's genus *Pentadenia* (1973, 1975, 1977, 1981, 1984) into sect. *Stygnanthe.* My treatment expands the circumscription of sect. *Pentadenia*; other species, such as *C. oblongifolia* and *C. poortmannii*, not placed by Wiehler in his genus *Pentadenia*, are here included in sect. *Pentadenia.*

1. Columnea atahualpae J. F. Smith & L. E. Skog, Novon 3: 186. 1993.—TYPE: ECUADOR. El Oro: forest along trail from Sambotambo, following headwaters of Río Moro Moro S to Buenaventura, at and along hwy to Portovelo, 29 Aug 1943, *Steyermark 54228* (holotype: F!)

Suffrutescent terrestrial (possibly epiphytic) herbs with erect stems to 1 m tall, 8–12 mm in diameter, terete, tawny-brown, proximally glabrous, distally appressed-pubescent with golden-yellow uniseriate trichomes; internodes 2.5–9.0 cm long; nodes slightly swollen; leaf scars slightly raised. Leaves opposite, isophyllous to slightly anisophyllous; laminas 9.2–22.5 cm long, 3.7–7.8 cm wide, obovate to elliptic, apex acute, base cuneate-oblique, adaxially green, strigose to appressed-pilose with single-celled transparent and uniseriate golden-yellow trichomes, abaxially silvery green-maroon, vesture similar to that of adaxial surface but sparser, veins abaxially sericeous, lateral veins 6–9, margin entire to slightly undulate; petioles 1.0–3.2 cm long, appressed-pilose with uniseriate golden-yellow trichomes. Inflorescences of 1–7 flowers in either axil of leaf pair; floral bracts 1–2 of unequal size, 8–22 mm long, 1.5–16 mm wide, lanceolate to ovate, apex acuminate, slightly constricted at base, reddish, appressed-pilose with uniseriate golden-yellow trichomes, margin entire. Pedicels 1.3–1.9 cm long, erect in leaf axil, appressed-pilose with uniseriate golden-yellow trichomes, glands 1 to several, ca. 0.4 mm long, purplish. Calyx clasping corolla at base but more open towards apex of calyx lobes; lobes equal, 1.5–2.7 cm long, 0.3–0.5 cm wide, lanceolate to subulate, apex long-acuminate, green with red-purple tips, exterior surface sericeous with uniseriate golden-yellow trichomes, inte-

FIG. 4. *Columnea atahualpae*. A. Habit. B. Adaxial leaf pubescence. C. Pedicel and calyx. D. Flower. E. Corolla interior with stamens. F. Gynoecium with nectaries. G. Ovary with calyx and nectaries. H. Berry with nectaries. I. Seeds. (A–B, D–F, based on *Steyermark 53640*; C, G–I, based on *Steyermark 54228*.)

rior similar but indumentum sparser, margin serrulate, 3–5 subulate teeth per side, mostly at base. Corolla 3.0–3.3 cm long, 5–8 mm wide at widest point, 4–5.5 mm wide before limb, 3.0 mm wide at base, tubular, gibbous at base, slightly ventricose, pale yellow, exterior surface appressed-pilose to sericeous with uniseriate golden-yellow trichomes, interior pilose to hirsute proximally with transparent uniseriate trichomes, with capitate-glandular trichomes dorsally and distally; lobes equal, 2.0 mm long, 1.5–2.0 mm wide, semiorbicular to acute. Filaments connate at base for 5 mm, adnate to corolla tube base for 3 mm, white, slightly pubescent, curling after anthesis, included in corolla; anthers 1.7 mm long, 1.7 mm wide, quadrate, included in corolla tube. Ovary 3 mm long, sericeous; style white, proximally pilose, distally glandular; stigma bilobed, papillose, included in corolla tube. Nectary of 5 glands, the 2 dorsal ones enlarged. Berry 1.0 cm in diameter, globose, color unknown but probably pale pink, pubescent; seeds 1.2–1.3 mm long, fusiform, red-brown, twisted, striate. Fig. 4.

Distribution (Fig. 5). Ecuador; known only from a small geographic area in El Oro and one other collection in Zamora-Chinchipe; 1035–1890 m. The Zamora-Chinchipe collection is a poor specimen, and its placement in *C. atahualpae* is based on its appressed golden-yellow vesture. This collection may represent another species or perhaps *C. nervosa*.

ADDITIONAL SPECIMENS EXAMINED. **Ecuador.** EL ORO: vicinity of Ayapamba, *Rose & Rose 23461* (NY); 2 leagues NE of Curtincapa, bordering Quebradas Nudillo and Tambillo, tributary to Río San Luis and Piedra Grande, *Steyermark 53840* (F, US).—ZAMORA-CHINCHIPE: "Palandra" [Palanda?], *André 4688* (NY).

Columnea atahualpae is very similar to *C. nervosa*; however, the much larger obovate to elliptic leaves with entire margins, smaller corollas, and large floral bracts are sufficient to separate it from *C. nervosa*. The golden-yellow appressed trichomes, found throughout the plant, distinguish *C. atahualpae* from other species of *Columnea*. This species is probably closely related to *C. lophophora*, which also has clustered flowers, large floral bracts, and dense vesture. *Columnea atahualpae* can easily be distinguished from *C. lophophora* by its tightly appressed pubescence, which is neither long-sericeous nor as dense as in *C. lophophora*.

2. Columnea hypocyrtantha (Wiehler) J. F. Smith & L. E. Skog, Novon 3: 190. 1993. *Pentadenia hypocyrtantha* Wiehler, Phytologia 73: 234. 1992.—TYPE: BOLIVIA. Cochabamba: Prov. Chaparé, rd from Cochabamba to Santa Cruz, 2500 m, 15 Jan 1965, *Vogel 498* (holotype: US!; isotypes: A, F, Z).

Suffrutescent, sublignose, epiphytic herbs with upright ascending stems, to more than 50 cm tall, 4–5.5 mm in diameter, terete, tawny, proximally smooth and glabrous, distally slightly pilose; internodes 0.7–32 mm long; nodes slightly swollen; leaf scars raised. Leaves opposite, isophyllous to slightly anisophyllous; laminas 6.3–13.3 cm long, 2.9–8.1 cm wide, ovate, apex acuminate, base rounded, slightly oblique, adaxially dark green, slightly pilose with uniseriate transparent trichomes, abaxially maroon or green suffused with purple, glabrate to slightly pilose with uniseriate transparent trichomes on the veins, lateral veins 4–7 per side, margin minutely denticulate; petioles 0.9–2.1 cm long, slightly pilose with uniseriate transparent trichomes. Inflorescences of a solitary flower per axil; floral

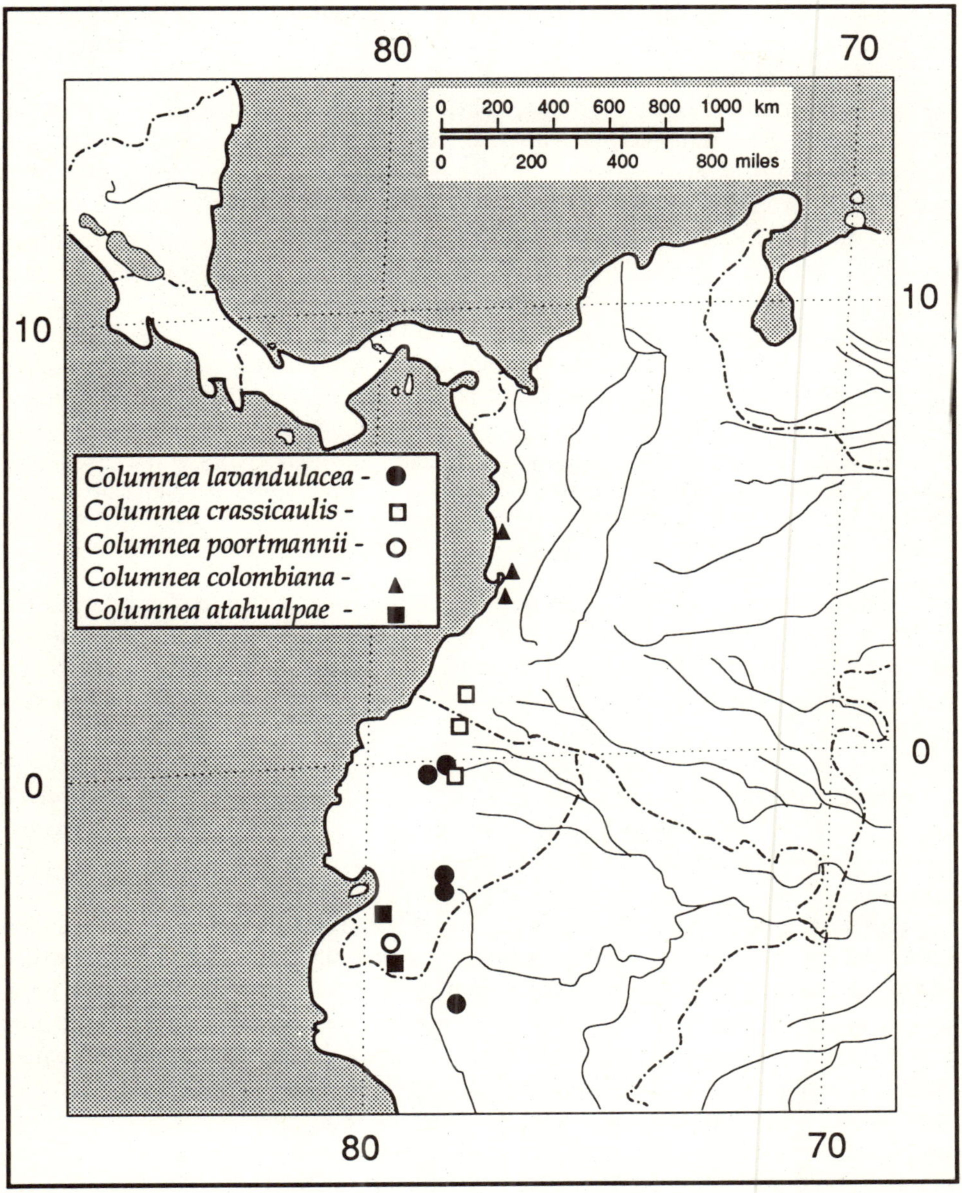

FIG. 5. Distribution of *Columnea lavandulacea, C. crassicaulis, C. poortmannii, C. colombiana,* and *C. atahualpae.*

bracts absent or perhaps caducous. Pedicels 4–6.4 cm long, erect in axil, red-purple, slightly pilose with uniseriate transparent trichomes, glands near calyx 0.5–0.7 mm long, oblong, dark in color. Calyx clasping corolla; lobes equal, 12–16 mm long, 4.5–9 mm wide, ovate, apex long-acuminate, green with red tips to maroon, both surfaces glabrate, margin entire to ciliate. Corolla 2.4–3.5 cm long, 1.1–1.8 cm wide at widest point, 3.5–4.5 mm wide before limb, 4–8 mm wide at base, tubular, strongly ventricose-saccate with pouch projecting forward beyond the opening of the corolla, strongly constricted at opening, slightly gibbous at base, orange-red, limb and lobes green, exterior surface glabrate to pubescent, more densely pubescent toward limb, interior glabrous; lobes equal, ca. 1 mm long, 2 mm wide, semiorbicular. Filaments connate at base for 5–9 mm, adnate to base of corolla tube for 2–3 mm, white, glabrous; anthers 1.5 mm long, 1.5 mm wide, quadrate, included in corolla tube. Ovary 6 mm long, glabrous; style white, glabrous; stigma stomatomorphic, smooth, included in corolla tube. Nectary of 5 free glands, 2 dorsal glands enlarged, slightly connate. Berry 9 mm in diameter, globose, glabrate, green with purple stripes; seeds not seen. Fig. 6.

Phenology. Collected only in January.

Distribution (Fig. 12). Bolivia (Cochabamba), known only from the type and a few collections nearby in the provinces Chaparé and Carrasco; 2000–2950 m.

ADDITIONAL SPECIMENS EXAMINED. **Bolivia.** COCHABAMBA: Prov. Carrasco, Km 262 on rd from Santa Cruz–Cochabamba, 2 km E of Siberia, *Besse et al. 1725* (US); Prov. Carrasco, Serrania Siberia, 20–35 km W of Comarapa on old Cochabamba–Santa Cruz rd (Hwy 4), *Dorr & Barnett 7045* (US); Prov. Chaparé, rd from Cochabamba to Villa Tunari, *Luer et al. 564* (SEL).

Columnea hypocyrtantha is similar to *C. trollii* to which it is presumedly related on the basis of its strongly ventricose corolla. It is easily distinguished from *C. trollii* by its glabrate, ovate leaves, glabrous ovary, and the corolla pouch, which projects forward beyond the opening of the corolla, a character not seen in any other species of *Columnea*.

3. Columnea isernii Cuatrecasas, Anales Ci. Univ. Madrid 4: 247. 1935. *Pentadenia isernii* (Cuatrecasas) Wiehler, Phytologia 27: 315. 1973.—TYPE: ECUADOR. Bolívar(?): El Jorge, *Isern 502* (holotype:MA; photo:MA!)

Suffrutescent, sublignose, epiphytic or terrestrial herbs with upright ascending stems to 2 m tall, 5–7 mm in diameter, terete, proximally smooth to scaly, glabrous, distally sericeous with uniseriate transparent trichomes; internodes 1–10.7 cm long, shorter internodes with leaves clustered at apex; nodes slightly swollen; leaf scars raised. Leaves opposite, generally clustered at apex, isophyllous to slightly anisophyllous; laminas 3.9–18.1 cm long, 1.8–6.2 cm wide, elliptic to elliptic-lanceolate, apex acute to acuminate, base oblique-cuneate, adaxially green to red-green, appressed hirsute with uniseriate, occasionally single-celled, transparent trichomes, abaxially pale green to maroon, appressed-hirsute or strigose with uniseriate or single-celled trichomes, veins abaxially sericeous with uniseriate trichomes, lateral veins 6–9, margin serrulate to crenate; petioles 0.4–3.6 cm, appressed-pilose to sericeous. Inflorescences of 1–5 flowers per leaf axil; floral bract 1, 7–10 mm long, 1.5 mm wide, frequently caducous, lanceolate, apex acuminate, red, sparsely pilose, margin entire. Pedicels 0.6–3.0 cm long, erect in leaf axil, villous to sericeous, eglandular. Calyx clasping corolla; lobes approximately equal, 7–20 mm long, 3–4 mm wide, ovate to lanceolate, apex blunt, red, sericeous at base tending to glabrous at apex on both sides, dissected at base, serrulate at apex. Corolla 1.8–3.2 cm long, 5–7 mm wide at widest point, 2–6 mm wide before limb, 3.5–4 mm wide at base, tubular, constricted at apex and at base, yellow-green and purple-

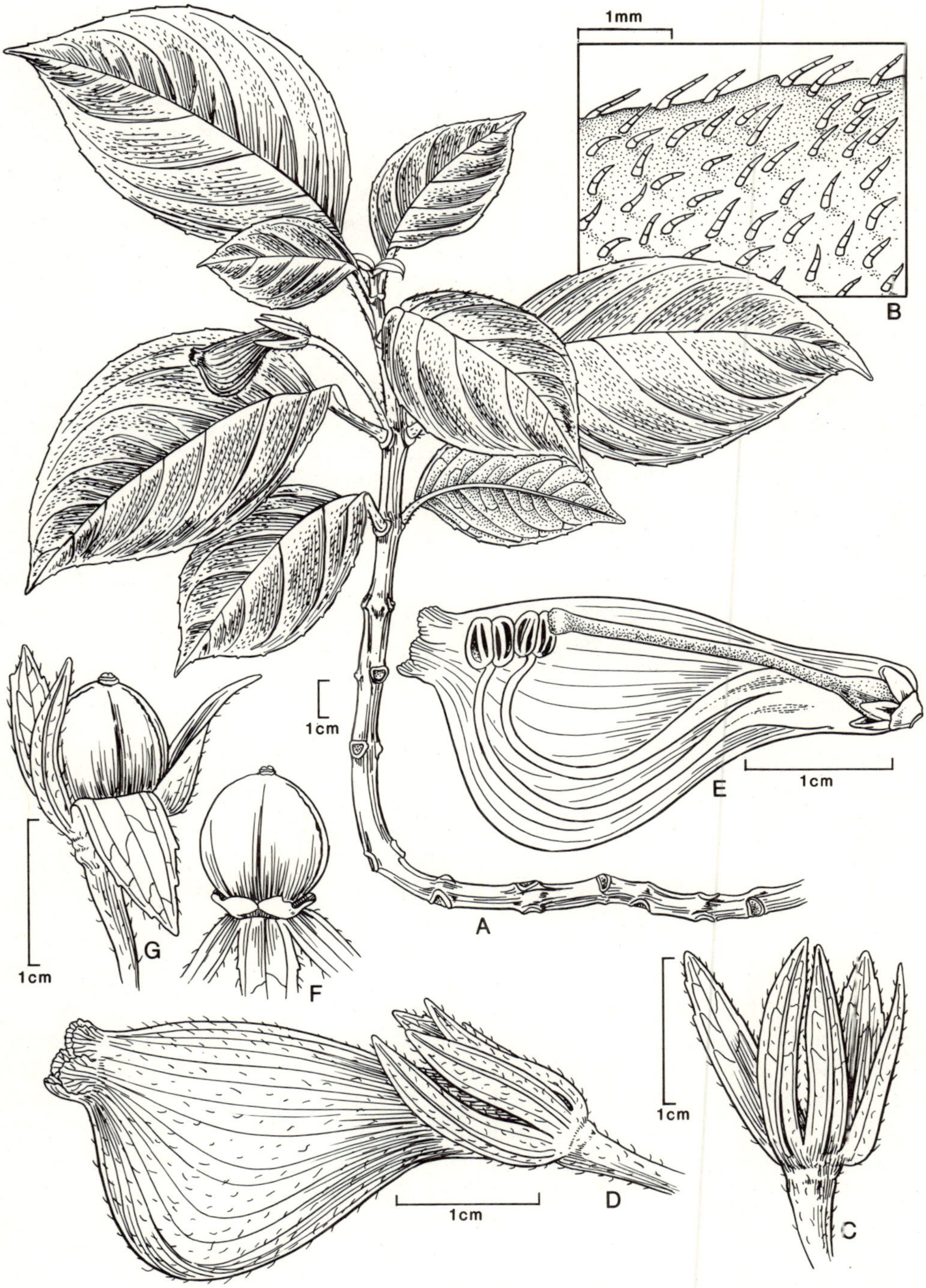

FIG. 6. *Columnea hypocyrtantha.* A. Habit. B. Abaxial leaf pubescence. C. Pedicel and calyx. D. Flower. E. Corolla interior with stamens. F. Berry with calyx lobes pulled away to show nectaries. G. Berry with calyx in natural position. (A–B, based on *Besse et al. 1725*; C–G, based on *Dorr & Barnett 7045*.)

striped, exterior surface villous to sericeous, interior pilose at base with capitate-glandular trichomes dorsally and distally; lobes equal, ca. 1.5 mm long, 1.5 mm wide, semiorbicular. Filaments connate at base for 3 mm, adnate to base of corolla tube for 2 mm, white, glabrous; anthers 1.5–1.7 mm long, 1.5–1.7 mm wide, quadrate, included in corolla tube. Ovary 4 mm long, glabrous to sparsely pubescent; style white, glabrous; stigma bilobed, papillose, included in corolla tube. Nectary of 5 free glands. Fruit and seeds not seen. Fig. 7.

Phenology. Flowering from November to January, possibly earlier or later in the season.

Distribution (Fig. 8). Ecuador (Bolívar, Cañar) and Peru (Piura, Tumbes); seasonally dry forests to cloud forests; 600–1600 m.

ADDITIONAL SPECIMENS EXAMINED. **Ecuador.** BOLÍVAR: Atio de Telimbela, *Acosta S. 6904* (F).—CAÑAR: Km 110 from Duran, *Dodson & Thien 2095* (WIS); rd from Huigra–Guayaquil, *Smith 2010* (QCA, QCNE, WIS, US).—LOJA: Celica–Alamor rd, at crossing with Río Alamor, *Harling & Andersson 22199* (US). **Peru.** PIURA: Prov. Huancabamba, Canchaque, "Chorro Blanco," *Díaz et al. 2775* (MO, WIS).—TUMBES: Prov. Zarumilla, Bosque Nacional de Tumbes, trail that connects Nito. Figueroa with El Caucho to Campe Verde, near Cerro de Tres Picos to rd, along rd to Campo Verde, *Simpson & Schunke 382* (F, G, US).

Columnea isernii is easily distinguished from other species of *Columnea* by its unusual yellow and purple-striped, densely sericeous corolla and by its calyx lobes, which are dissected at the base. Cladistic analyses of chloroplast DNA restriction site variation and morphology (Fig. 2) show that this species is closely related to *C. nervosa* (Smith & Sytsma 1994a, b, c).

4. Columnea lophophora Mansfeld, Biblioth. Bot. 116: 145. 1937. *Pentadenia lophophora* (Mansfeld) Wiehler, Phytologia 27: 315. 1973.—TYPE: ECUADOR. Chimborazo: above Huigra, Río Chanchan, *Diels 1177* (B, destroyed).—ECUADOR. Chimborazo: close to Huigra, mostly on Hacienda de Licay, 21 Aug 1918, *Rose & Rose 22281* (neotype, designated by Kvist & Skog, 1993: US!; isoneotype: NY!).

Suffrutescent, sublignose, terrestrial or epiphytic herbs with stems to 3.0 m tall, 8–11 mm in diameter, terete, tan, proximally smooth and glabrous, distally densely sericeous to appressed-villous with transparent uniseriate trichomes; internodes 0.2–6.5 cm long, the proximal larger, the distal shorter; nodes slightly swollen; leaf scars raised. Leaves opposite, isophyllous to slightly anisophyllous, clustered at apex of the stem; laminas 4.7–12.5 cm long, 2.6–4.6 cm wide, ovate to elliptic, apex acute, base rounded, adaxially green, densely sericeous with uniseriate transparent trichomes, abaxially silvery white-pink, densely sericeous to lanate with transparent or reddish uniseriate trichomes, veins abaxially sericeous with longer hairs than lamina, lateral veins ca. 7 per side, margin serrulate to denticulate, with red uniseriate trichomes; petioles 0.3–1.8 cm long, densely sericeous with long uniseriate transparent trichomes. Inflorescences of 3–6 axillary flowers (possibly more); each inflorescence subtended by 3 bracts, the center the largest, 15–32 mm long, 7–15 mm wide, ovate to elliptic, acuminate, slightly constricted at the base, green-pink, densely sericeous to villous, trichomes longer on veins, margin entire, 3–5 lateral veins per side; each flower subtended by 2 bracts, 1.5 cm long, 1.0 cm

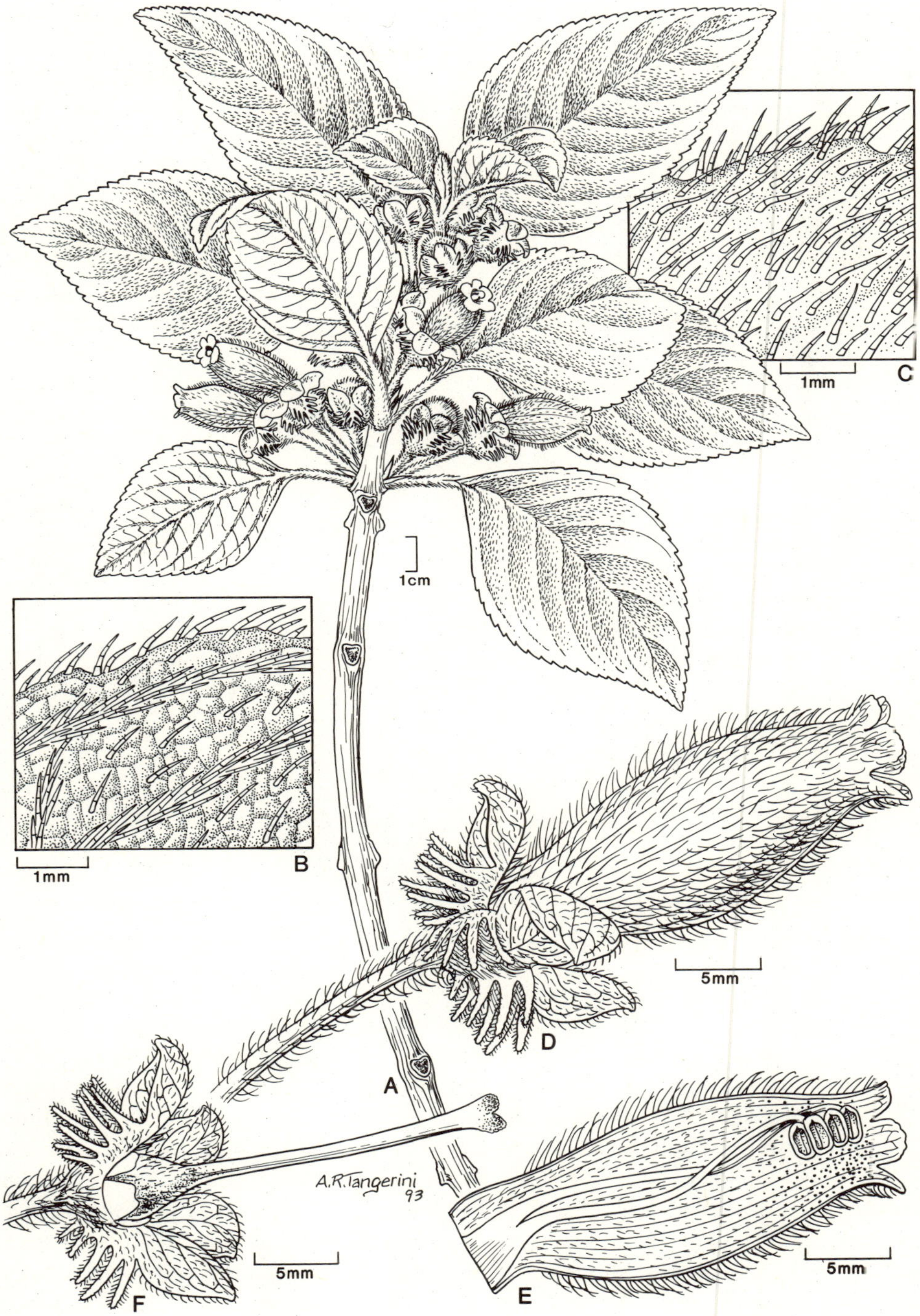

FIG. 7. *Columnea isernii*. A. Habit. B. Abaxial leaf pubescence. C. Adaxial leaf pubescence. D. Flower. E. Corolla interior with stamens. F. Gynoecium with nectaries and calyx. (A, C–F, based on *Dodson & Thien 2095*; B, based on *Harling & Andersson 22199*.)

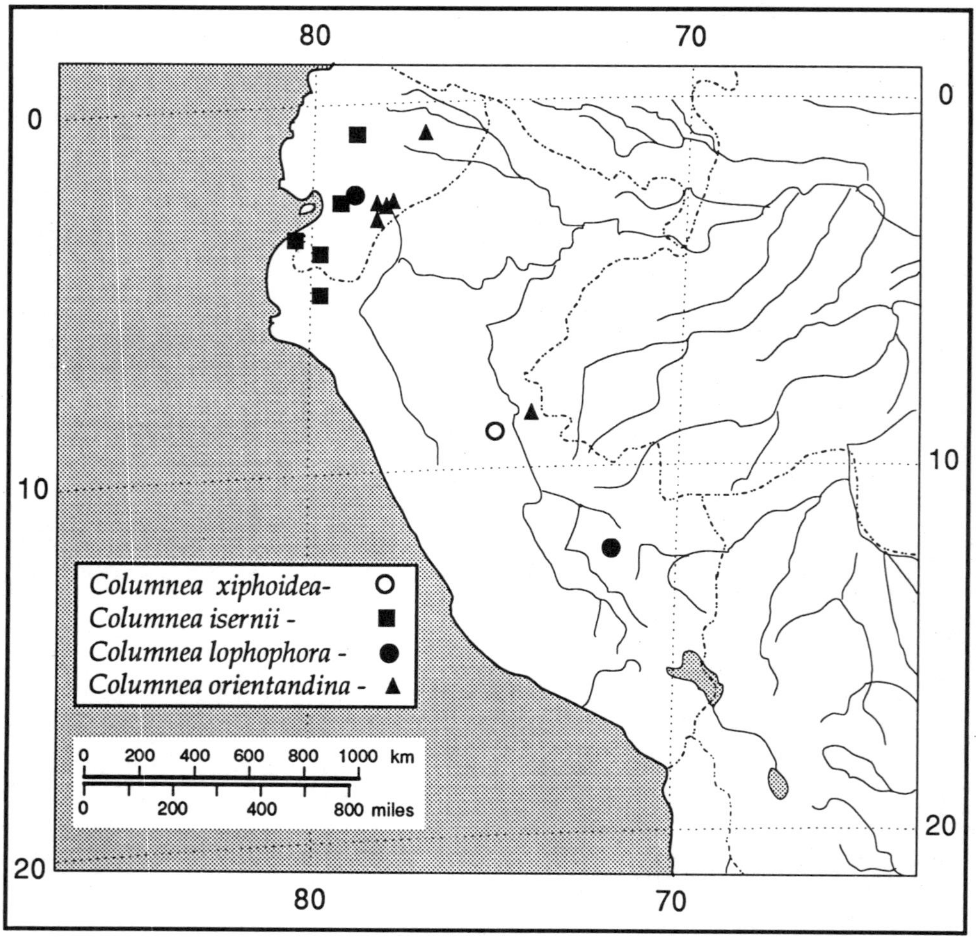

FIG. 8. Distribution of *Columnea xiphoidea, C. isernii, C. lophophora,* and *C. orientandina.*

wide, ovate to elliptic, apex abruptly acuminate, densely sericeous. Pedicels absent or up to 4.5 cm long, erect in the axils, densely sericeous, eglandular. Calyx clasping corolla; lobes equal to subequal, 11–23 mm long, 2.5–7.0 mm wide, lanceolate, apex acute, green with pink tips to pink, densely sericeous with both pink or transparent uniseriate trichomes, serrulate with few teeth at apex or entire. Corolla 2.9–3.5 cm long, 6–12 mm wide at widest point, 5–7 mm wide before limb, 4–8 mm wide at base, tubular, slightly ventricose, violet-purple to blue, exterior surface densely sericeous, becoming even more densely pubescent toward limb, interior sericeous at base, gradually becoming glabrous toward mouth, capitate-glandular trichomes dorsally toward mouth; lobes equal, 2–3 mm long, 2–3.5 mm wide, semiorbicular. Filaments connate at base for 3–5 mm, adnate to base of corolla tube for 1–2 mm, slightly pilose; anthers 1.7 mm long, 1.7 mm wide, quadrate, included in corolla tube. Ovary 4–6 mm long, pilose to sericeous; style reddish, slightly pilose near ovary or glabrous; stigma bilobed, smooth, included in corolla tube. Nectary of 5 free glands, the dorsal 2 enlarged, or of only 2 enlarged dorsal

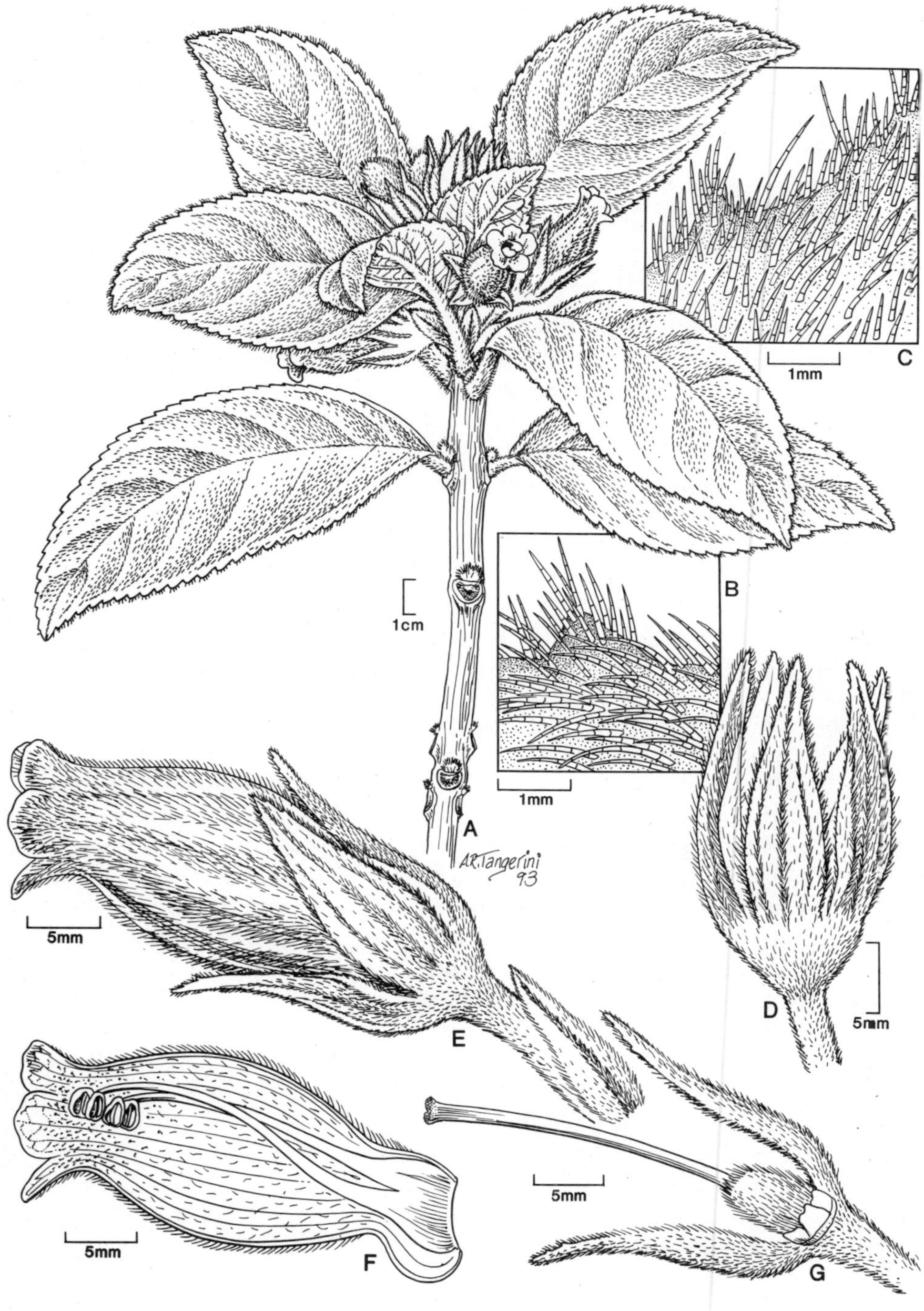

FIG. 9. *Columnea lophophora*. A. Habit. B. Abaxial leaf pubescence. C. Adaxial leaf pubescence. D. Pedicel and calyx. E. Flower. F. Corolla interior with stamens. G. Gynoecium with nectaries (three calyx lobes removed). (Based on *Iltis & Ugent 1055*.)

glands. Only immature berries seen, these globose, pubescent, and of dark color when dried; seeds not seen. Fig. 9.

Distribution (Fig. 8). Ecuador and Peru (Cuzco); known from one seasonally dry region and one wet-forest region in southern Peru; 1500–2300 m.

ADDITIONAL SPECIMENS EXAMINED. **Ecuador.** Locality unknown: *Spruce 5529* (NY, P, W). **Peru.** CUZCO: Urubamba, Sanctuario Histórico Machu Picchu, *Iltis et al. 1055* (US); Urubamba, Pampachua, 95 km from Cuzco, between Km 95 and 107 on trail from Cuzco to Quillabamba, *Núñez V. & Bengoa 8551* (MO, US, WIS); Urubamba, Sanctuario Histórico Machu Picchu, *Vargas C. 18335* (US).

Columnea lophophora is a rare species, previously believed to be a local endemic to Ecuador prior to the discovery of the Peruvian localities (Kvist & Skog 1993). Although the Peruvian specimens are in rather poor condition (the corollas are not attached, and the floral bracts are absent), they are undoubtedly *C. lophophora*. The dense vesture and unusual corolla color distinguish *C. lophophora* from any other species of *Columnea*.

Columnea lophophora is most closely related to *C. nervosa* and *C. isernii* as indicated by cladistic analysis of morphological characters (Smith & Sytsma 1994b). The apically clustered leaves and alternation of regions of long and short internodes in this species, and others related to it, seem to indicate a habitat that is seasonally dry resulting in a corresponding slower growth rate (Kvist & Skog 1993).

The species may be extinct in Ecuador, as suggested by Kvist and Skog (1993). During field work in 1989, I relocated Hacienda de Licay. It lies in the hills above Huigra (north), and could be reached at that time by following the railroad tracks west out of town and taking the first trail on the right leading into the hills. The native vegetation was almost entirely destroyed and replaced with grassy, weedy vegetation and a few, occasional agaves. Occasional remnants of native vegetation consisted of species more frequently associated with wetter habitats, such as species of *Begonia* and *Anthurium*. Perhaps the original vegetation, composed mainly of species adapted to drier habitats, included occasional pockets of wetter vegetation types. *Columnea lophophora* may have existed in some of these pockets. Beyond the summit of the hill at Hacienda de Licay is a small quebrada that also has species associated with moister habitats, such as species of *Passiflora, Begonia, Calceolaria,* Loasaceae, and *Heppiella ulmifolia* (H. B. K.) Hanst. (Gesneriaceae). Although most of these species are indicators of moister habitats, they are also tolerant of disturbance. *Columnea lophophora* was not found during this trip; few species of sect. *Pentadenia* are tolerant of human disturbance, and it is possible that *C. lophophora* is now extinct in Ecuador. Yet, recent explorations among isolated patches of native vegetation in this area have revealed populations of *Fuchsia insignis* Hemsl., another species previously believed to be extinct (P. Berry, pers. comm.); perhaps populations of *C. lophophora* may yet be discovered in Ecuador.

5. Columnea nervosa (Klotzsch ex Oersted) Hanstein, Linnaea 34: 400. 1865. *Pentadenia nervosa* Klotzsch ex Oersted, Centr. Gesn. 57. 1858.—TYPE: [PANAMA.] "Veragua," *Warscewicz s.n.* (holotype: B, destroyed).—PANAMA. Veraguas: southwestern slope of Volcán Barú, summer 1968, *Butcher s.n.* (neotype, here designated: US!).

Alloplectus rubidus C. Morton, Ann. Missouri Bot. Gard. 24: 204. 1937. *Columnea rubida* (C. Morton) C. Morton, Baileya 7: 58. 1959. *Pentadenia rubida* (C. Morton) Wiehler, Selbyana 2: 122. 1977.—TYPE: PANAMA. Chiriquí: Valley of upper Río Chiriquí Viejo, vicinity Monte Lirio, 27 Jun–13 Jul 1935, *Seibert 141* (holotype: US!; isotypes: F! K! GH! MO! NY!).

Columnea vinacea C. Morton, Contr. U.S. Natl. Herb. 29: 38. 1944. *Alloplectus vinaceus* (C. Morton) D. Gibson, Phytologia 23: 334. 1972. *Pentadenia vinacea* (C. Morton) Wiehler, Phytologia 27: 315. 1973.—TYPE: GUATEMALA. Sololá: Volcán Atitlán, south facing slopes, 11 Jun 1942, *Steyermark 47408* (holotype: US!; isotype: F!).

Pentadenia matudae Wiehler, Selbyana 2: 122. 1977. *Columnea matudae* (Wiehler) L. P. Kvist & L. E. Skog, Allertonia 6: 391. 1993.—TYPE: MEXICO. Chiapas: Rodillo, 9 Aug 1937, *Matuda 1626* (holotype: US!; isotypes: F! GH! K! MO! MEXU, NY!).

Suffrutescent, epiphytic or terrestrial herbs, sparsely branching with ascending, sometimes scandent sublignose stem up to 1 m tall, 4–10 mm (possibly greater) in diameter at base, terete, maroon or green, proximally glabrous or nearly so, distally appressed villous to sericeous; internodes 0.2–10.3 cm long and generally appearing in alternating clusters of short and long internodes throughout stems; nodes swollen; leaf scars raised. Leaves opposite, generally clustered at the apex of the stem, isophyllous to slightly anisophyllous; laminas 3.3–18.0 cm long, 1.5–10.2 cm wide, ovate to elliptic, apex acute or acuminate, base cuneate-oblique and decurrent, adaxially green to maroon, appressed-pilose to hirsute with uniseriate transparent hairs, occasionally red, abaxially green to maroon, frequently with red-purple mottling, strigose with single-celled transparent hairs, sericeous on veins with uniseriate hairs, lateral veins 5–9 per side, margin serrulate to denticulate, vesture on margin composed of red-purple, uniseriate trichomes; petioles 0.4–4.0 cm long, green, occasionally red, sericeous with transparent uniseriate trichomes, rarely red with red trichomes running into midvein. Inflorescences of 1–4 axillary flowers in either axil of a leaf pair; floral bract 1, 3–13 mm long, 1–2.5 mm wide, occasionally caducous, lanceolate, green or red, sericeous. Pedicels 0.8–3.7 cm long, erect, sericeous with transparent uniseriate trichomes, red in specimens with redder leaves, glands at base of calyx frequently obscured by vesture, 1.0–1.5 mm long, oblong to fusiform, red-purple. Calyx clasping corolla; lobes equal, 8–28 mm long, 1–6 mm wide, lanceolate, subulate, green, or with red tips, or entirely red to maroon, sericeous both surfaces, less densely sericeous towards apex and on interior surface, margin serrate or rarely entire, 2–5 subulate teeth per side, mostly near the base. Corolla 2.5–6.0 cm long, 6–14 mm wide at widest point, 3–8 mm wide before limb, 2.5–6.0 mm wide at base, tubular and slightly ventricose, gibbous at base, red with yellow-green limb and limb lobes, or entirely yellow, exterior surface sericeous with long red or transparent uniseriate trichomes, interior less densely sericeous and with capitate-glandular trichomes distally and dorsally; lobes equal to unequal, 1.5–5 mm long, 3.2–5 mm wide, semiorbicular, lower lobes occasionally narrow and reflexed. Filaments connate and adnate to base of corolla tube for 4 mm, white, glabrous; anthers 1–1.7 mm long, 0.8–1.7 mm wide, subquadrate, included in corolla tube. Ovary 3–8 mm long, sericeous with uniseriate transparent trichomes; style white or occasionally red, generally glabrous but occasionally with a few uniseriate transparent trichomes proximally and capitate-glandular trichomes distally; stigma bilobed or stomatomorphic, white, smooth, rarely exserted. Berry 4–8 mm in diameter, globose, white, sericeous; seeds to 1.1 mm long, oblong, twisted, shiny red-brown, striate.

Phenology. Flowering from April to August, fruiting from July to December, but infrequently collected in flower during other times (flowering may be continuous). Populations from Ecuador have a longer period of flowering, extending into December and starting as early as March.

Distribution (Fig. 10). Southern Mexico, Guatemala, El Salvador, Panama, and Ecuador; wet forests; 1250–3800 m, generally below 2500 m.

ADDITIONAL SPECIMENS EXAMINED. **Mexico.** CHIAPAS: above El Rosario, 8 mi S of Motozintla, *Croat 40734* (MO); Mpio. Angel Albino Corzo, El Triunfo, *MacDougall 466* (BH, US [2]); Mt. Ovando, *Matuda 1860* (GH, US), *17743* (US); in wet forest, San Juan Panamá, Escuintla, *Matuda 19070* (F). **Guatemala.** JALAPA: Volcán Jumay, N of Jalapa, *Steyermark 32327* (F, US).—QUETZALTENANGO: 3 km S of Santa María Planta Electrica on Hwy 9s, *Roe et al. 723* (WIS).—SAN MARCOS: Barranca Eminencia, above San Rafael Pie de la Cuesta, *Standley 68656* (F); Barranca Eminencia, rd between San Marcos and San Rafael Pie de la Cuesta, in upper part of barranca between Finca La Lucha and Buena Vista, *Standley 86312* (F).—SUCHITEPÉQUEZ: Volcán Santa Clara, between Finca El Naranjo and upper slopes, *Steyermark 46698* (F, US).—Department unknown: Santa María, *Bernoulli & Cario 2010* (GOET). **El Salvador.** AHUACHAPÁN: Cerro Grande de Apaneca, Sierra de Apaneca, *Weberling 2630* (WAG).

Ecuador. LOJA: Catacocha–Macará rd, ca. 16 km N of Macará, *Harling & Andersson 15313* (US); between Utuana and Colaisaca, *Harling et al. 20703* (GB, US); 8 km W of Celica, *Harling & Andersson 22156* (US); Celica–Alamor rd, ca. 3 km W of Celica, *Harling & Andersson 22176* (US); San Bartolo, *Poortmann 501* (K, P); 4.8 km W of Celica, *Smith 1963* (QCA, QCNE, US, WIS).—ZAMORA-

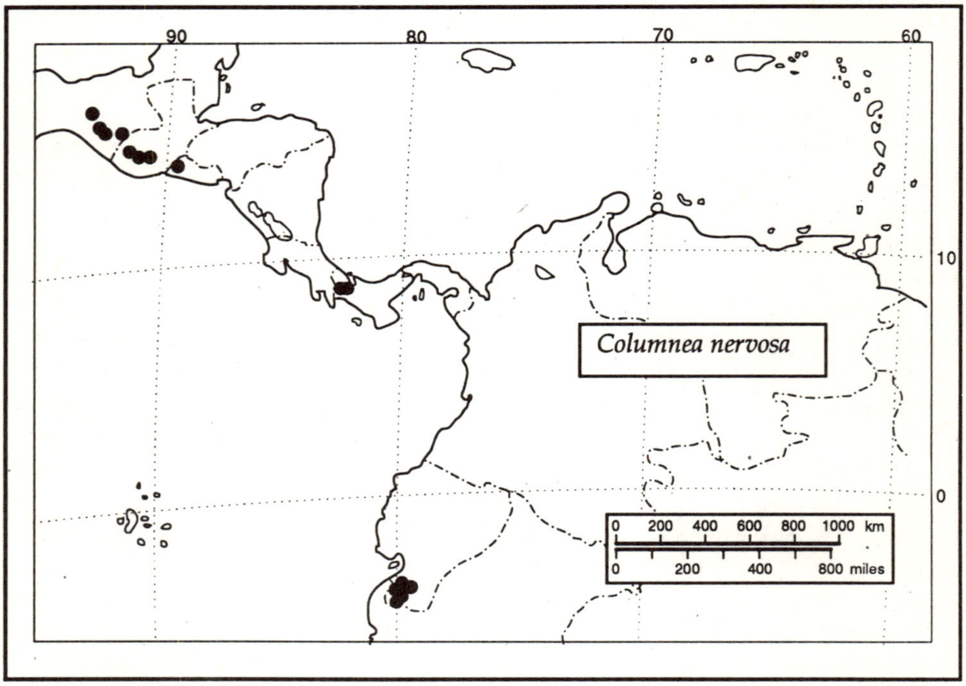

FIG. 10. Distribution of *Columnea nervosa*.

CHINCHIPE: Savanella, *Poortmann 383* (K).—Province unknown: *Poortmann 207* (P); Cerro de Santa Rosa, *Poortmann 243* (P).

Columnea nervosa is characterized by its shrubby habit, leaves generally clustered at the apex of the stems, small to caducous floral bracts, and mid-sized to large corollas (2.5–6.0 cm long). *Columnea isernii* differs by its yellow and purple-striped corollas and *C. poortmannii* by its more ventricose corollas with a narrower opening. The larger floral bracts of *C. atahualpae* distinguish it from *C. nervosa*. *Columnea lophophora* is separated by its much denser pubescence and large conspicuous floral bracts.

The alternation of short and long internodes along the length of the stem and clustering of leaves at the apex in *C. nervosa* is a likely indicator of a seasonal habitat (Mansfeld 1937; Kvist & Skog 1993). The habitat in which it occurs is a transition between evergreen and semi-deciduous forest in Mexico (Rzedowski 1981), and semi-deciduous forest in Ecuador (Harling 1979). It is likely that the internodes lengthen appreciably in the wet season, but grow very little during the dry period. Flowering occurs in the dry season, and flowers are generally found along the regions of the shorter internodes.

Columnea nervosa, as recognized here, has been described as four different species, but there is little morphological variation to distinguish among the segregates. In general, the Central American collections differ from those of Ecuador only in having all-yellow corollas and redder vegetative structures. Ecuadorian plants are generally larger with longer and wider corollas, longer calyx lobes, and larger and more ovate leaves; however, the variation seen in the collections gathered outside of Ecuador represents a subset of the variation seen within Ecuador. The Central American populations are likely the result of long-distance dispersal from Ecuador. Variation in corolla color within a species is apparently common in *Columnea* and is also seen in other species, such as *C. angustata*, *C. spathulata*, *C. strigosa,* and *C. oblongifolia*.

Cladistic analyses of morphology, chloroplast DNA restriction site variation, and the combination of these two data sets all indicate that *C. nervosa* is closely allied to *C. isernii* (Fig. 2) (Smith & Sytsma 1994a, b, c).

See the discussion of *C. poortmannii* (no. 7) concerning the confusion of type material of *C. poortmannii* and specimens of *C. nervosa*.

6. Columnea oblongifolia Rusby, Mem. Torrey Bot. Club 6: 98. 1896. *Ortholoma oblongifolium* (Rusby) Wiehler, Phytologia 27: 322. 1973. *Trichantha oblongifolia* (Rusby) Wiehler, Selbyana 1: 35. 1975.—TYPE: BOLIVIA. La Paz: "Mapiri," Jan 1893, *Bang 1744* (holotype: NY; isotypes: BM, E, F, GH-2! K-2! M, MANCH, MO! US-2! W, WU).

Shrubby, sublignose, epiphytic or terrestrial herbs with upright or rarely pendent stems to 1 m, 2.5–7 mm in diameter, terete, red-brown, proximally scaly and glabrous, distally pubescent to glabrate; internodes 0.4–12 cm long; nodes swollen; leaf scars raised. Leaves opposite, slightly anisophyllous; laminas 3.5–14 cm long, 2.5–9.3 cm wide, oblong to elliptic, apex acute to acuminate, base rounded or cuneate, oblique, adaxially green, hirsute with uniseriate transparent trichomes, abaxially green to purple, strigose with single-celled transparent trichomes, veins appressed hirsute with uniseriate transparent trichomes, lateral veins 6–9 per side; margin serrulate (rarely serrate); petioles 4–15 mm long, appressed-hirsute with uniseriate transparent or red trichomes. Inflorescences each of 1 flower in either axil of a leaf pair; floral bract 1, 7 mm long, 2 mm wide, caducous, green. Pedicels 2.9–9.0 cm long, erect to pendent, hirsute to sericeous with uniseriate transparent trichomes, glands conspicuous, 1.5 mm long, fusiform, dark purple. Calyx loosely clasping corolla; lobes equal to subequal, 10–23 mm long, 1–3 mm wide, dorsal lobes slightly reflexed, linear-lanceolate, apex acuminate, green, margin entire, exterior surface hirsute, interior strigose to pilose. Corolla 4.0–6.5 cm long, 1.4–2.3 cm wide at widest point, 1.0–1.6 cm wide before limb, 5–8 mm wide at base, tubular, ventricose, constricted at base and slightly at limb, lateral and lower lobes reflexed, upper 2 lobes fused into a galea, red-purple with yellow limb and lobes, or all-yellow, exterior surface hirsute, interior glabrate to glandular, hirsute at opening and on adaxial surface of lobes; lobes equal, 3–4 mm long, 3–10 mm wide, semiorbicular. Filaments connate at base for 4–5 mm, adnate to base of corolla tube for 1–2 mm, white, glabrous; anthers 1.5–2.0 mm long, 2 mm wide, subquadrate, slightly exserted beyond mouth of corolla. Ovary 5 mm long, tomentose with uniseriate transparent or red trichomes; style white, glandular; stigma stomatomorphic, smooth, included in corolla tube. Nectary of 2 dorsal connate glands. Berry 1.2 cm in diameter, globose, rose-violet, tomentose; seeds 1–1.1 mm long, fusiform, twisted, red-brown, striate. Fig. 11.

Phenology. Flowering from October to March, possibly earlier and later, fruiting in April.

Distribution (Fig. 12). Southern Peru to northern Bolivia; wet forests and cloud forests; 2000–2900 m.

Additional Specimens Examined. **Peru.** Ayacucho: Prov. Huanta, *Weberbauer 7568* (US).—Cuzco: Prov. Urubamba, Sanctuario Histórico Machu Picchu, top of Huayna Picchu, by cave, *Iltis & Ugent 1059* (WIS); Urubamba, hillside of Puncuyoj, Sanctuario Histórico Machu Picchu, 10 km SW of Incatambo, *Peyton & Tilney Peyton 1382* (MO); La Convención, Huayopata, Puncuyoj in Protipata drainage, 10 km SW of Inca Tambo, *Peyton & King 1405* (US); Sanctuario Histórico Machu Picchu, 0.5 km N of union of Sayacmarca and Atobamba rivers, *Peyton and Tilney Peyton 1488* (MO, US); Sanctuario Histórico Machu Picchu, Inca Trail past Inti Puncu, *Smith 1721* (CUZ, USM, WIS); Waynapijchu, *Vargas C.17109* (US); Prov. Calca, cañón de Lo'ak'ay, *Vargas C. 23283* (US).—Puno: Prov. Carabayo, Ollacha, *Vargas C. 6942* (US). **Bolivia.** La Paz: Prov. Murillo: 27.4 km N of dam at Lago Zongo, *Solomon 8973* (MO); 30.5 km N of dam at Lago Zongo, trail up Río Jachcha Cruz, *Solomon 9053* (MO, WIS). Prov. Nor Yungas: Chuspipata via Coroico, *Beck 8772* (US); 11 km NE of Chuspipata on Coroico rd, *Gentry & Solomon 44723* (MO); 4.5 km below Yolosa, then ca. 23 km W on rd up to Río Huarinilla, *Solomon 8864* (MO); 11 km from Chuspipata to Yolosa, *Smith & Beck 1725* (LPB, WIS).

The large red-purple or yellow corollas of *C. oblongifolia* distinguish this species from other species of *Columnea*. It is similar in overall appearance, both vegetatively and reproductively, to *C. strigosa*, which has reflexed corolla lobes and an orange corolla that is more constricted at the base of the limb.

Wiehler placed *C. oblongifolia* in the genus *Trichantha* (=sect. *Ortholoma*), but its robust, shrubby habit, and ventricose corolla indicate that it correctly belongs in sect. *Pentadenia*. Morphological and cpDNA variation show it is most closely related to *C. trollii* (Fig. 2) (Smith & Sytsma 1994a, b, c). *Columnea oblongifolia* is similar to *C. trollii* in that the fleshy indehiscent berry, typical of

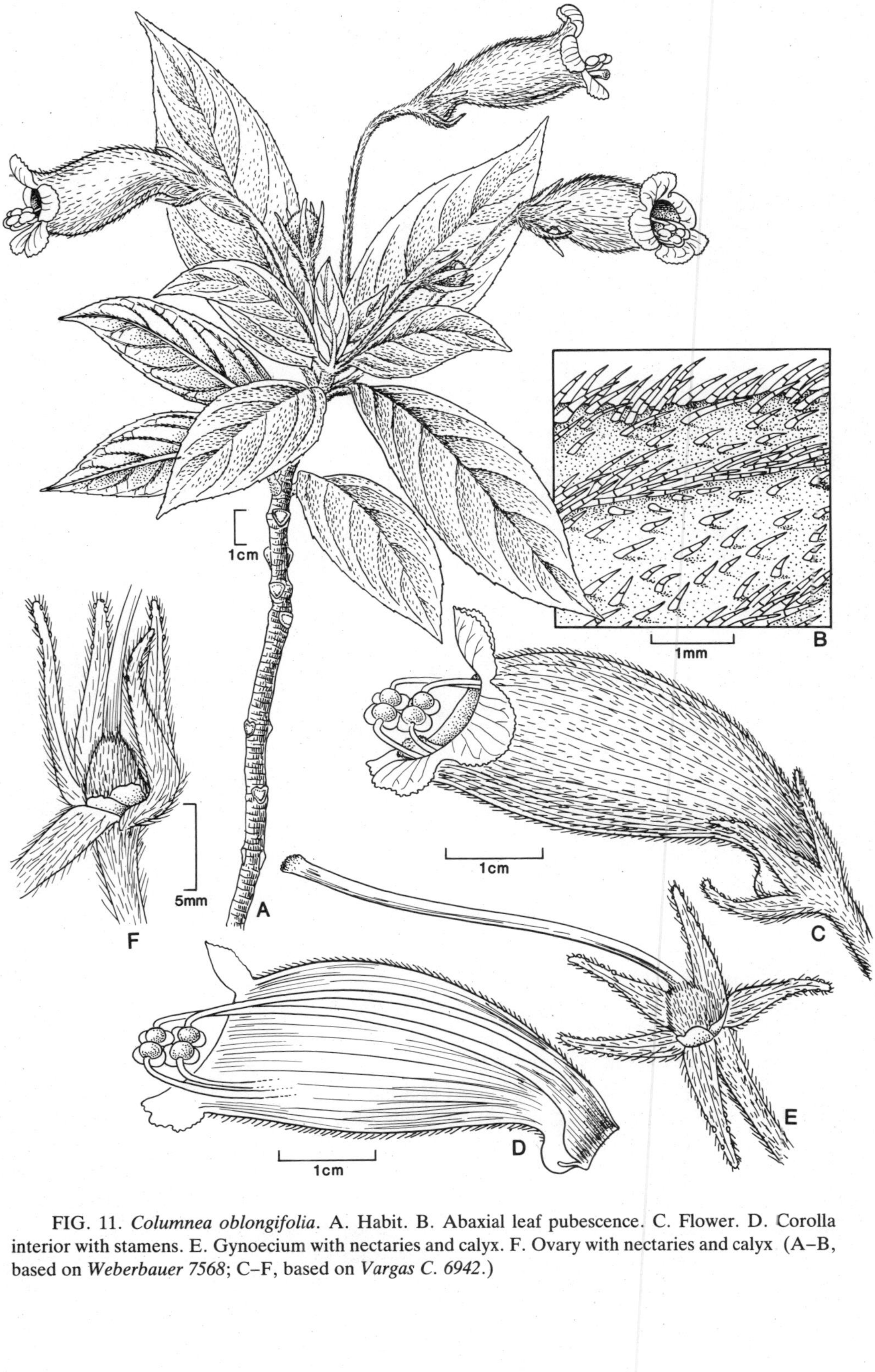

FIG. 11. *Columnea oblongifolia*. A. Habit. B. Abaxial leaf pubescence. C. Flower. D. Corolla interior with stamens. E. Gynoecium with nectaries and calyx. F. Ovary with nectaries and calyx (A–B, based on *Weberbauer 7568*; C–F, based on *Vargas C. 6942*.)

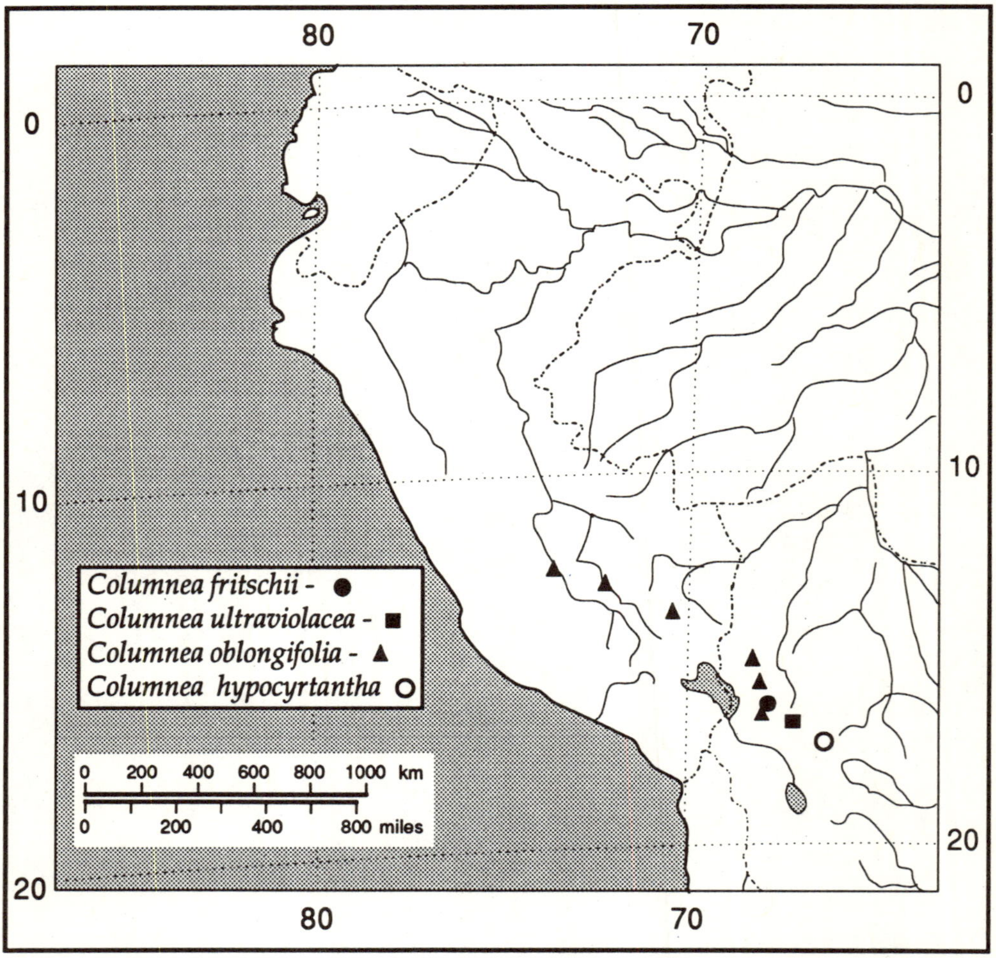

FIG. 12. Distribution of *Columnea fritschii, C. ultraviolacea, C. oblongifolia,* and *C. hypocyrtantha*.

Columnea, upon drying may split open along loculicidal lines, dividing the fruit in half. *Columnea oblongifolia*, like *C. trollii*, exhibits a character state for fruit dehiscence intermediate between *Columnea* and *Alloplectus*.

7. Columnea poortmannii (Wiehler) L. P. Kvist & L. E. Skog, Allertonia 6: 394. 1992. *Trichantha poortmannii* Wiehler, Selbyana 7: 340. 1984.—TYPE: ECUADOR. Loja: Cordillera de Zamora, near Loja, 27 Dec 1881, *Poortmann 265* (holotype: P!; isotype: P!).

Suffrutescent herbs with erect stems to 15 cm tall, 3–5 mm in diameter (probably greater), terete, red-brown, proximally smooth, distally sericeous; internodes 2–42 mm long; nodes swollen; leaf scars raised. Leaves opposite, isophyllous to slightly anisophyllous, clustered at apex of stems; laminas 3.3–6.2 cm long, 2.2–3.5 cm wide, ovate, apex acute to acuminate, base cuneate-oblique, adaxially green, villous to lanate with transparent or red uniseriate trichomes, abaxially green, hirsute with transparent uniseriate trichomes, lateral veins 5–7 per side; margin serrulate; petioles 0.5–1.8 cm long, lanate with transparent uniseriate trichomes. Inflorescences each of 1 flower in either axil of a leaf pair; floral bracts absent or possibly caducous. Pedicels 1.2–2.4 cm long, erect, reddish, hirsute with transparent uniseriate trichomes, eglandular. Calyx clasping corolla; lobes unequal, 14–22 mm long, 2–5 mm wide, elliptic to lanceolate, apex acute, green-reddish or with red tips, exterior surface hirsute to villous, interior glabrous with a few trichomes near apex, margin serrate. Corolla 4.5–5.7 cm long, 9–14 mm wide at widest point, 3–5 mm wide before limb, 3–4 mm wide at base, tubular, gibbous at base, strongly ventricose and sharply constricted at the throat, probably pale red or yellow, with pale green limb, exterior surface pilose, interior with fewer trichomes but with capitate-glandular trichomes dorsally and distally; lobes equal, ca. 2.5 mm long, 2.5 mm wide, semiorbicular. Filaments connate at base for 3–4 mm, adnate to base of corolla tube for 1–2 mm, white, slightly pubescent; anthers 1.5 mm long, 1.5 mm wide, quadrate, included in corolla tube. Ovary 4.5 cm long, glabrous or nearly so; style white, with few glandular trichomes; stigma stomatomorphic, white, smooth, included in corolla tube. Nectary composed of 2 glands. Only immature berries seen, these ovoid, dark-colored, densely hirsute; seeds not seen.

Distribution (Fig. 5). Southeastern Ecuador; 2300 m.

ADDITIONAL SPECIMEN EXAMINED. **Ecuador.** LOJA: Zamora–Huaico, *Espinosa 2259* (US).

Columnea poortmannii is apparently a rare endemic from Ecuador, collected only once since the type material. It differs from other species of *Columnea* by its strongly ventricose corolla and densely hirsute to lanate leaves. In appearance, the corolla is similar in size and shape to that of *C. trollii*; however, these species are easily distinguished by their sharply different corolla structure and leaf pubescence. In addition, leaves of *C. trollii* are elliptic, and those of *C. poortmannii* are ovate.

In the protologue of *C. poortmannii*, Wiehler lists the type as *Poortmann 265*. He annotated two specimens of this collection at P as holotype and isotype of *C. poortmannii*; however, he also labeled a specimen of *C. nervosa*, *Poortmann 243* (P), as the holotype of *C. poortmannii*. Although the two species are quite distinct, the general appearance of the herbarium sheets for these two collections are very similar, and the designation of *Poortmann 243* (*C. nervosa*) as the type for *C. poortmannii* is clearly an error.

8. Columnea strigosa Bentham, Pl. Hartw. 232. 1846. *Pentadenia strigosa* (Bentham) Hanstein, Linnaea 26: 211. 1854.—TYPE: ECUADOR. Tungurahua: Mt. Tungurahua, *Hartweg 1262* (holotype: K!; isotypes: BM, CGE, E, G, NY! P! W).

Columnea campanulata Bentham, Pl. Hartw. 232. 1846.—TYPE: COLOMBIA. Cundinamarca: Bogotá, *Hartweg s.n.* (holotype: K!).

Columnea macrantha Bentham, Pl. Hartw. 232. 1846.—TYPE: ECUADOR. Pichincha: Andes near Quito, *Hartweg s.n.* (holotype: K!).

Columnea aurantiaca Decaisne ex Planchon, Fl. Serres Jard. Eur. 6: 45, pl. 552. 1850–1851. *Pentadenia aurantiaca* (Decaisne ex Planchon) Hanstein,

Linnaea 26: 211. 1854.—TYPE: VENEZUELA. Mérida: *Linden 1454* (holotype: MPU; isotypes: BM, G, GH! K! P! W-2!).

Columnea pichinchensis Hanstein, Linnaea 34: 398. 1865.—TYPE: ECUADOR. Pichincha: Quito, *Jameson 667* (holotype: LE; isotypes: BM, K! NY! P! US! W).

Columnea campanulata var. *longipedunculata* Cuatrecasas, Trab. Mus. Nac. Ci. Nat., Ser. Bot. 33: 121. 1936.—TYPE: COLOMBIA. Tolima: Ibagué, *Cuatrecasas 2431* (holotype: MA; photo: MA!).

Columnea kucyniakii Raymond, Svensk. Bot. Tidskr. 58: 185. 1969.—TYPE: cultivated material grown from seedlings collected in Cañar, Ecuador, *Raymond s.n.* (holotype: MTJB!).

Herbaceous, epiphytic or terrestrial shrubs with vining, scandent, or climbing stems 1–2 m long, 2–10 mm in diameter, terete, red-brown or tan, proximally smooth and glabrous, distally densely hirsute to glabrescent; internodes 0.2–14.7 cm long; nodes swollen; leaf scars raised. Leaves opposite, slightly anisophyllous; laminas 2.3–16.7 cm long, 1.2–13.7 cm wide, ovate to elliptic, rarely lanceolate, rarely large ovate-orbicular, apex acute to acuminate, base rounded or cuneate, oblique, adaxially green, strigose with single-celled transparent trichomes, rarely hirsute, abaxially green to purple, strigose to glabrous, veins hirsute, strigose, or glabrate, lateral veins 6–9 per side; margin entire to serrulate; petioles 0.2–6.0 cm long, strigose to hirsute. Inflorescences usually each of 1 flower in either axil of a leaf pair or rarely of 2–3 (–6) flowers; floral bract 1, 3–3.5 mm long, rarely up to 7 mm long, 0.5 mm wide, caducous, linear, red or green, hirsute. Pedicels 1.7–29.8 cm long, erect or the longer ones pendent, densely hirsute or strigose, glands conspicuous, near calyx, oval, red-purple. Calyx clasping corolla; lobes equal to subequal, 5–40 mm long, 3–12 mm wide, ovate to lanceolate, enlarged in fruit, apex acute to acuminate, long-acuminate in fruit, red or green, exterior surface strigose to glabrate, occasionally hirsute, interior glabrous, margin serrate at base. Corolla 2.8–9.5 cm long, 8–28 mm wide at widest point, 6–18 mm wide before limb, 3–12 mm wide at base, tubular, strongly ventricose, constricted at base and before limb, orange, rarely red-orange with yellow limb; 2 dorsal lobes fused completely or nearly so, forming a galea extending well beyond the opening of the corolla tube, exterior surface hirsute, interior sparsely hirsute to glabrous, frequently with glandular trichomes distally and dorsally; galea 4–21 mm long, 4.5–22 mm wide, lateral lobes 5–20 mm long, 6–20 mm wide, linear to ovate, slightly reflexed, ventral lobe 7–22 mm long, 2–16 mm wide, linear, slightly to strongly reflexed. Filaments connate at base for 3–4 mm, adnate to base of corolla tube for 1–2 mm, white, glabrous; anthers 1.5–5.5 mm long, 1.0–3.5 mm wide, rectangular, exserted. Ovary 3.5–8 mm long, villous to sericeous; style red, glabrous; stigma stomatomorphic, red, smooth, exserted. Nectary variable, of 5 free glands to only 2 dorsal connate glands. Berry 0.7–1.9 cm in diameter, globose, purple, hirsute to pilose; seeds 1–1.3 mm long, fusiform, twisted, red-brown, striate. Figs. 13, 14.

Phenology. Flowering and fruiting periods vary: in Venezuela, flowering from May to August and fruiting from October to November; in eastern and central Colombia, flowering from January to July (rarely into October) and fruiting from August to January; in southern Colombia and Ecuador, flowering continuously.

Distribution (Fig. 15). Western Venezuela to northern Peru; wet forests and cloud forests; 1500–3800 m.

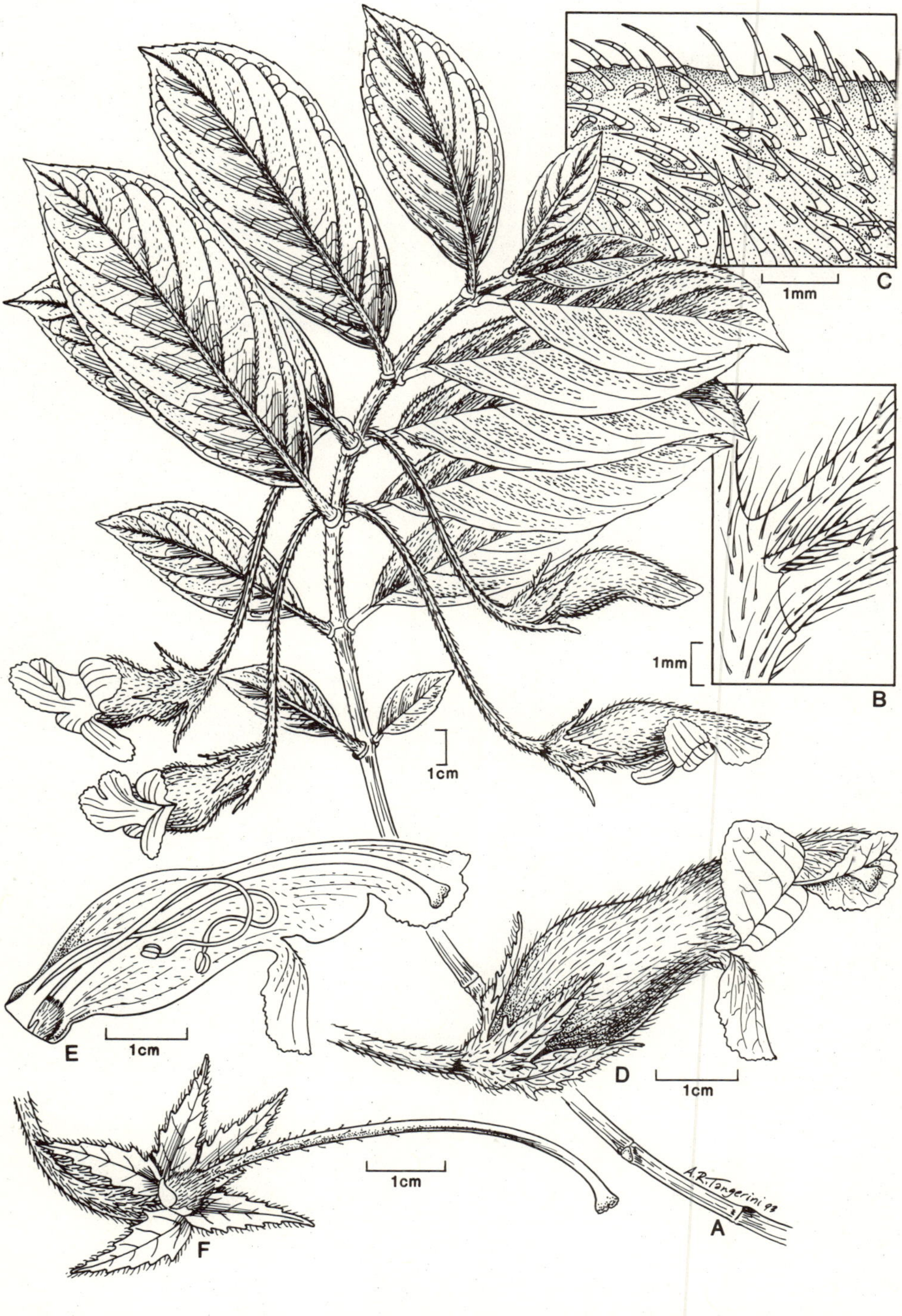

FIG. 13. *Columnea strigosa*. A. Habit. B. Bract at base of pedicel. C. Abaxial leaf pubescence. D. Flower. E. Corolla interior with stamens and gynoecium. F. Gynoecium with nectaries and calyx. (Based on *Smith & Díaz 1651*.)

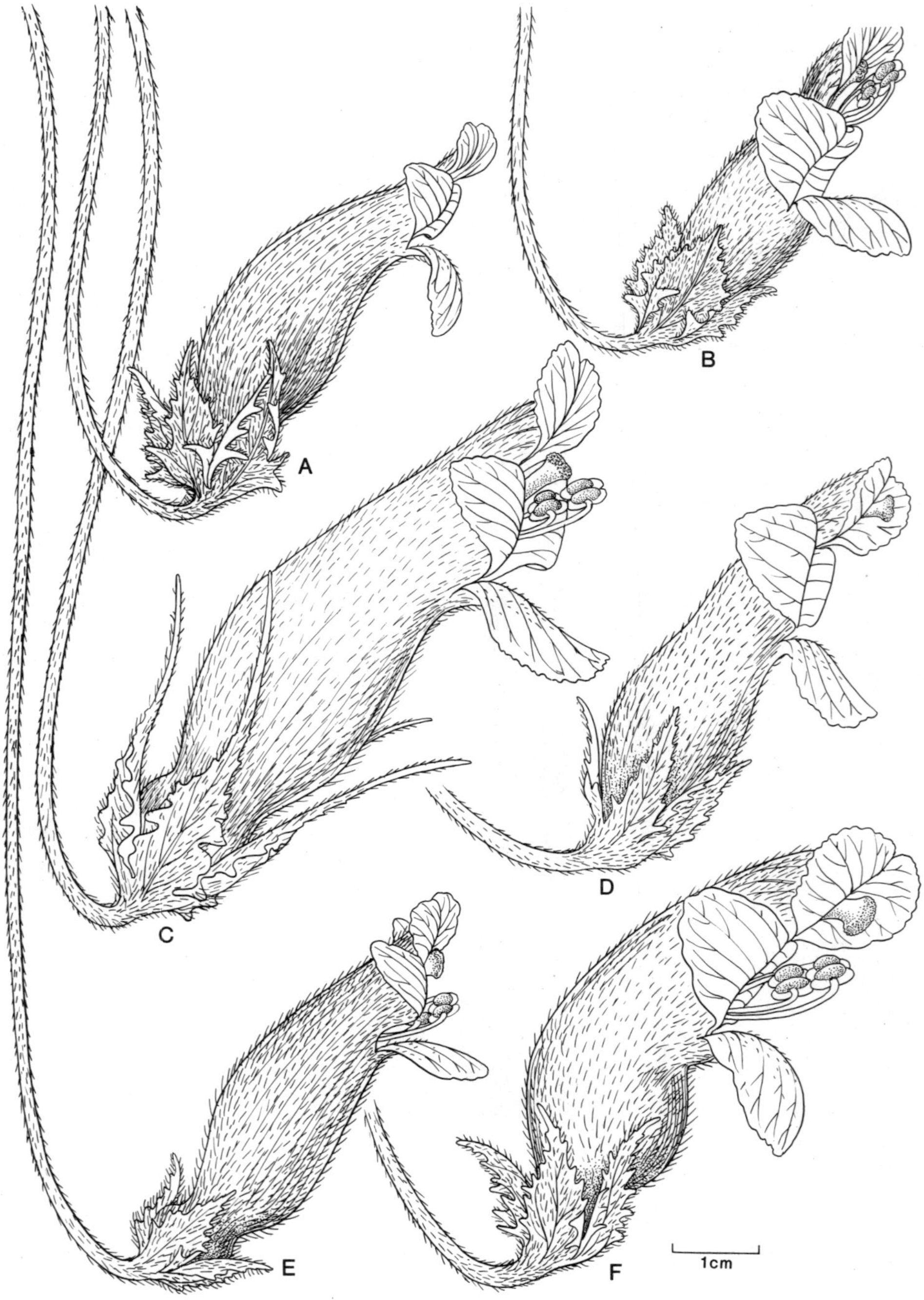

FIG. 14A–F. Morphological variation in flowers of *Columnea strigosa*.

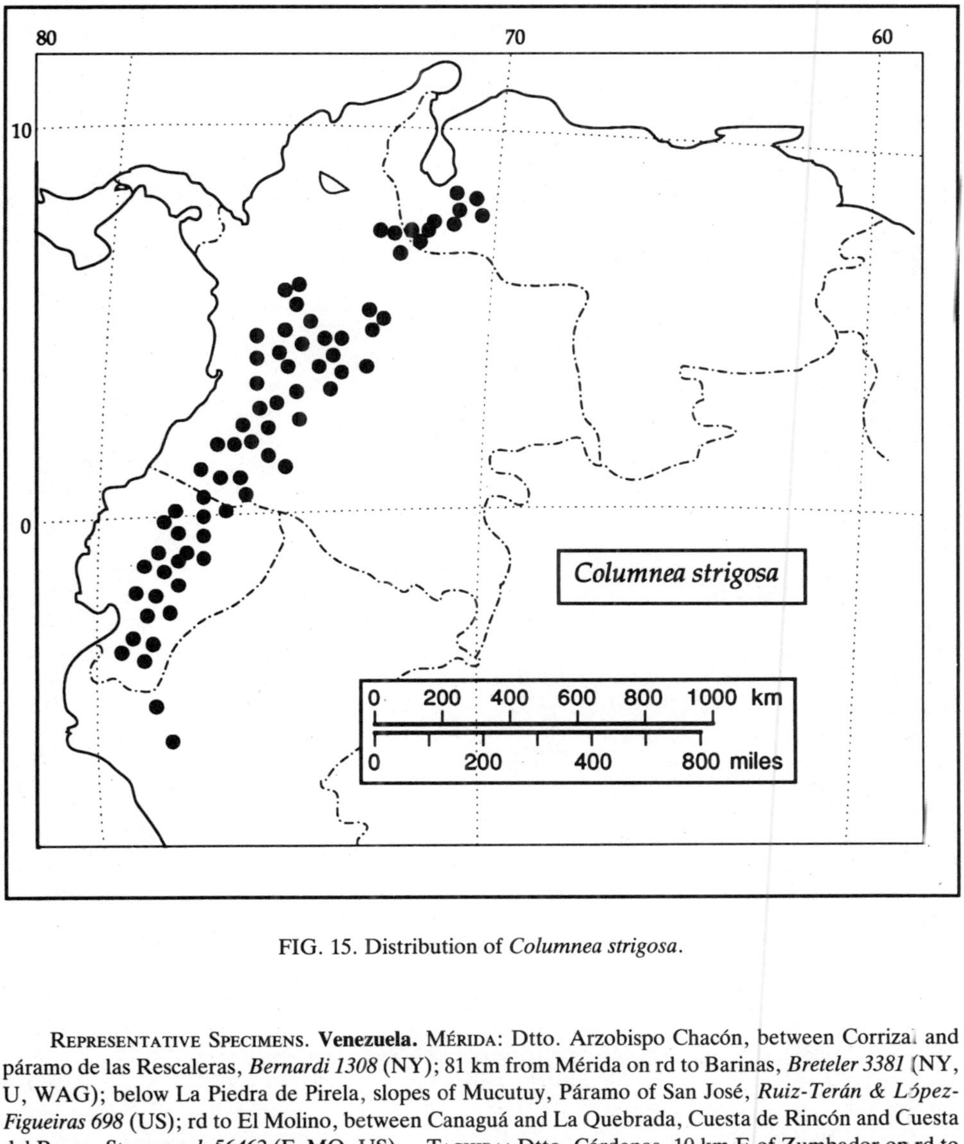

FIG. 15. Distribution of *Columnea strigosa*.

REPRESENTATIVE SPECIMENS. **Venezuela.** MÉRIDA: Dtto. Arzobispo Chacón, between Corrizal and páramo de las Rescaleras, *Bernardi 1308* (NY); 81 km from Mérida on rd to Barinas, *Breteler 3381* (NY, U, WAG); below La Piedra de Pirela, slopes of Mucutuy, Páramo of San José, *Ruiz-Terán & López-Figueiras 698* (US); rd to El Molino, between Canaguá and La Quebrada, Cuesta de Rincón and Cuesta del Barro, *Steyermark 56462* (F, MO, US).—TACHIRA: Dtto. Cárdenas, 10 km E of Zumbador on rd to Queniquea, *Berry 3419* (MO); slopes below Páramo de Tamá, *Luteyn et al. 6010* (NY, SEL); 6 km on rd to Pregonero past rd to La Grita, *Smith 1220* (WIS); Dtto. Junín, *Ruiz-Terán & López-Figueiras 8911* (US); base of Páramo de Tamá, between Betania and Tamá, *Steyermark 57189* (F, US), *Steyermark 57171* (US). **Colombia.** ANTIOQUIA: Medellín, *Docell s.n.* (W); 10 km N of Sonsón, *Guttiérrez V. et al. 18c164* (US); Mpio. Sonsón, Km 10 on rd Sonsón–Argelia, Páramos de las Papas or de Sonsón, *Londoño 406* (US); 10 km E of Sonsón, *Scolnik et al. 19An200* (COL, F, US); 7–15 km E of Sonsón on rd to Nariño, *Stein & Congollo 3565* (COL, MO, US).—BOYACA: Páramo de Bijagual, *Espinal T. & Montenegro M. 1346* (COL); Sierra Nevada del Cocuy, below El Playón, *Grubb et al. 814* (COL, US); Mpio. Pajarito, *Lozano C. et al. 819* (COL); near border of Santander, Finca "La Sierra," *Lozano C. 2557* (COL); Hacienda de la Rusia, below Páramo, *Uribe U. 4240* (COL).—CALDAS: "Alaska," above Salento, *Pennell 9384* (GH).—CAQUETÁ: Quebrada del Río Hacha, below Gabinete, *Cuatrecasas 5845* (F, US).—CAUCA: La Cumbre near Pico de Aguila, W of Morales, *Core 1103* (US); Popayán, *García-Barriga 4550* (COL); Páramo de las Papas, *Idrobo et al. 3033* (COL); Mpio. Puracé, Parque Nacional Natural de Puracé, *Lozano et al. 4498* (COL); S of "Llano," *Pennell 6957* (NY); "Canaan," Mt. Puracé,

Pennell & Killip 6608 (NY, US); headwaters of Río López, Río Palo basin, *Pittier s.n.* (US).—CHOCÓ: plant cultivated near Medellín, collected near El Carmen, on rd between Medellín and Quibdo, *Luer s.n.* (SEL).—CUNDINAMARCA: Páramo de Choáchi, *Acosta-A. et al. 585* (COL, U); 5.5 km W of junction of main rd from Bogotá to Fusagasugá to Salto Tequendama, *Croat 55445* (MO [2], US); San Miguel, *Cuatrecasas 6682* (COL); Páramo de Guasca, *Cuatrecasas 9468* (F, NY, US); Mpio. of La Calera, rd to Choáchi, *Díaz P. & Rangel 2232* (COL); Tequendama Falls, *Ewan & Schiefer 499* (GH); San Miguel, rd to Fusagasugá, *García-Barriga 12672* (COL [2], US); 3 km SW of Charquita, 25 km WSW of Bogotá, *Grant 9018* (US, WIS); Cerro Negro, Hato Grande, 15 km E of Gachetá, *Grant 9408* (US, WIS); Muchindote Valley, E side of Quebrada Negra, 13 km NE of Gachetá, *Grant 9590* (BH, COL, US, WIS); rd E from Guasca, *Haught 5822* (COL, US); rd Facatativa–Anolaima, *Haught 6215* (US); Mpio. of Sibaté, above San Miguel, *Huertas & Camargo 5884* (COL); SW of Sibaté, El Peñón, *Pennell 2423* (F, GH, MO, US); Salto Tequendama, *Schneider 301-A* (COL); Carupa, Peña de Sumangá, *Uribe U. 5911* (COL).—HUILA: pass between Garzón and Florencia, *Mason 13926* (COL,US); "Balsillas," on Río Balsillas, *Rusby & Pennell 823* (NY); between La Plata and Popayán, *Bogner 857* (BH, M).—META: Sumapaz, valley of Quebrada El Buque, *Díaz P. et al. 2827* (COL); Páramo de Sumapaz, *Fosberg 20945* (NY, US); Mpio. Quetamel, rd to Calvario, *Lozano C. 4044* (COL).—NARIÑO: Páramo del Buey, SE of Pasto, *Espinosa 3085* (COL, NY); 5 km N of Victoria, *Ewan 16175* (BH, US); ridge N of Sibundoy, *Folsom & Channell 10654* (US); Mpio. San Francisco, rd San Francisco–Nocos, *Mora 1376* (COL); Páramo del Tabano, near Pasto, *Uribe U. 5339* (COL).—NORTE DE SANTANDER: 6 km NW of Gramalote, 35 km W of Cúcuta, *Fosberg & Fassett 21727* (US); Ocaña, *Schlim 560* (P).—PUTUMAYO: beyond Sibundoy, *Foster & Foster 2001* (A); Páramo El Tabano, *García-Barriga 4550* (US); Páramo El Tabano, *García-Barriga 7801* (US), *García-Barriga 7851* (COL, US); Páramo de San Antonio, between La Laguna and La Cocha and Valley of Sibundoy, *Schultes 3217* (GH, US); Páramo de Tambillo, NE of valley of Sibundoy, *Schultes & Smith 3123* (COL, F, GH, US, WIS).—SANTANDER: rd Sta. Rosita–Onzaga, near S. Isidro and border of Boyacá, *Cleef 9847* (COL, U).—TOLIMA: Mpio. Sta. Isabel, near la Bodega, *Díaz P. & Jaramillo M. 2194* (COL); 11–13 km W of Fresno on rd from Mariquita to Manizales, *Stein 3549* (COL, MO, US).—VALLE DEL CAUCA: valley of Bugalagrande, between Las Azules and Las Violetas, *Cuatrecasas 20807* (F [2], U, US); valley of Río Cali, near Peñas Blancas, *López-F. 8335* (US [2]). **Ecuador.** AZUAY: 1–8 km N of Sevilla de Oro, *Camp E-4627* (NY); Sigsig-Gualaquiza, Camp. Molón, *Harling et al. 8210* (GB); Sevilla de Oro, *Harling et al. 8471* (GB, SEL); El Chorro, 6 km above Molleturo on rd to Cuenca, *Harling & Andersson 22861* (US).—BOLÍVAR: Atio de Pucará, rd from Telimbela, *Acosta S. 6982* (F); Atio de Telimbela, *Acosta S. 7152* (F); Simiatug, Hacienda Talajua, *Penland & Summers 572* (F, GH, US).—CAÑAR: between Rivera and Pindilig, *Holm-Nielsen et al. 29241* (K).—CARCHI: Km 58 Tulcán–Maldonado rd, *Boeke 860* (NY); rd Tulcán–Maldonado, *Harling & Andersson 12408* (GB); Camp. Machines, rd Tulcán–Maldonado, 10 km SE of Maldonado, *Harling & Andersson 12329* (GB [2]); trail Pun–Chingual, *Mexia 7615* (GH, US).—CHIMBORAZO: Hacienda "La Carmela," Sibambe, *Acosta S. 5455* (F); rd Pusucucho–Placer, *Acosta S. 7250* (F); rd Baños–Puela, 2–3 km N of Manzano, *Lugo S. 1319* (GB, WIS).—COTOPAXI: Pilaló, trail N of village, *Boeke 539* (NY[2], US); Km 82 Quevedo–Latacunga, *Dodson & Dodson 15421* (MO).—IMBABURA: "Shanshipamba," La Esperanza, *Acosta S. 14408* (F).—LOJA: 55 km rd from San Lucas, *Dodson & Thien 744* (US); rd Alamor–Cazedoros, W of El Limo, *Harling & Andersson 22285* (US); 6 km SW of Loja along footpath, *Knight 132* (WIS); near Las Juntas, *Rose et al. 23187* (NY, US).—MORONA-SANTIAGO: Camp. San Miguel, rd Sigsig–Gualaquiza, *Harling et al. 8050* (GB); rd Limón–Gualaceo, Km 16–18 from Limón, *Harling & Andersson 12703* (GB); Pica Sevilla de Oro–Méndez, between Páramo del Castillo, Cerro Negro, and Dominguillos, *Jaramillo 5537* (MO).—NAPO: 1.1 km E of Cuyujúa, at Río Victoria, *MacBryde & Dwyer 1275* (US); Km 195 Lago Agrio–Quito rd, between Cuyujúa and Papallacta, *Stein 3083* (US).—NAPO-PASTAZA: Santa Bárbara de Sucumbios, *Harling 4081* (S).—PICHINCHA: near Quito, *Asplund 10080* (S); below San Juan towards Chiriboga, *Asplund 18983* (S); 6–20 km E of Tandapi, rd Quito–Santo Domingo, *Besse et al. 180* (SEL); Km 46 rd Quito–Santo Domingo, *Dodson & Thien 1045* (US); Tandayapa–Nono, near Tandayapa, *Harling & Andersson 11606* (GB); SE of El Playón de San Francisco, slopes of Cerro Mirador, *Holm-Nielsen et al. 29748* (K), *Holmgren 621* (S); 9 km SE of Cosanga between Cosanga and Tena, *Kirkbride & Chambra R. 4121* (US); Km 32–38 old rd Quito–Santo Domingo, *Luteyn & Luteyn 5607* (K, MO, NY, US); old rd Quito–Santo Domingo, *Smith 1849* (QCA, QCNE, WIS); rd Quito–Tandayapa, *Smith 1927* (QCA, QCNE, US, WIS), *Smith 1953* (BH, QCA, QCNE, US, WIS); rd Nono–Nanegal, Km 43–45, *Sparre 15981* (S); Nono–Nanegalito rd, 9 km from Nono, *Stein 2856* (MO, US).—SANTIAGO-ZAMORA: between La Esperanza and Sta. Ana, Huamboya, *Acosta S. 7404* (F); SE of El Pan, above Camp. Campanas,

Steyermark 53519 (F, US); 10 leagues SE of El Pan, between Campanas and Arenillas, *Steyermark 53540* (F).—TUNGURAHUA: N slopes of Tungurahua, Chaupi, *Dodson & Thien 2042* (BH, WIS); Colonia Regina, 5–6 km NW Río Verde, *Lugo S. 1912* (GB); below Minza páramo, *Penland & Summers 415* (F, GH, US); Mt. Tungurahua, *Rimbach 493* (S).—ZAMORA-CHINCHIPE: Nudo de Sabanilla to Valladolid, *Harling et al. 20474* (GB); above Valladolid on rd to Yangana, *Harling & Andersson 21440* (US); rd Loja–Zamora, Km 24–25, *Holm-Nielsen et al. 3463* (NY, S); new rd Loja–Zamora, *van der Werff & Palacios 9010A* (MO). **Peru.** AMAZONAS: Bagua, Cordillera Colón, SE of La Peca, *Barbour 3688* (MO, US); Bagua, Cordillera Colón, SE of La Peca, *Barbour 4114* (MO).—CAJAMARCA: Prov. Cutervo, La Pucarilla (Sócota–San Andrés), *López et al. 5389* (US); La Pucarilla (Sócota–San Andrés), *López et al. 6607* (US); Parque Nacional de Cutervo, *Smith & Díaz 1600* (CUZ, US, USM, WIS), *Smith & Díaz 1651* (BH, MO, QCA, SEL, US, USM [2], WIS).—Department unknown: *André 2092* (F, GH, NY), *Lobb 60* (K, W)

Although variable, *C. strigosa* is easily distinguished from other species of *Columnea.* In sect. *Pentadenia*, it is easily identified by its large, orange, bilabiate corollas. Although species in sect. *Columnea* also have bilabiate corollas, none possess the strongly ventricose corollas of *C. strigosa*. Morphologically, *C. strigosa* is most similar to *C. oblongifolia*, of southern Peru and Bolivia, which has bright yellow corollas with smaller lobes.

Columnea strigosa is a morphologically variable species, and many of the variants have been recognized at the species level. Most of the variation in *C. strigosa* is found in the shape of the corolla (Fig. 14). *Columnea macrantha* has somewhat larger, much more ventricose corollas with proportionally smaller lobes and a narrower constriction at the limb, only one flower per leaf axil, and oblong leaves. *Columnea pichinchensis* is the opposite, with a shorter, narrower corolla tube and proportionally larger lobes, slightly more ovate leaves, and an inflorescence of 2–3 flowers per axil. *Columnea kucyniakii* represents a morphological extreme with very large, ovate-orbicular leaves, up to 6 flowers per inflorescence, and corollas that are smaller, narrower, and with purple spots at the sinuses of the lobes. Kvist and Skog (1993) attributed the differences between *C. pichinchensis/kucyniakii* and *C. strigosa* to environmental differences caused by an either epiphytic or terrestrial habitat. Other species of *Columnea* occasionally also exhibit such variance, and specimens assigned to *C. pichinchensis* and *C. kucyniakii* were both terrestrial and epiphytic, as were specimens of *C. strigosa.* Kvist and Skog (1993) also proposed that these taxa may have been at one time distinct species, and that introgression has since reduced the distinctions between them to a continuum of variation.

It is also possible that full speciation has not yet occurred within this complex. The morphological variation, seen in the species as a whole, may be the result of numerous alleles in low frequency. In some populations, one of these alleles may rise in frequency to the point that it becomes locally common. A single subtle difference may not merit taxonomic recognition, but numerous such differences may cause the elevation of those populations to species status.

The morphological variants in *C. strigosa* may be due to the fixation, or high frequency, of an allele controlling a particular morphological feature. Thus, some populations may be noticeably distinct but not to such a degree to merit specific status; *C. pichinchensis* and *C. kucyniakii* may represent such populations. Examination of the entire spectrum of morphological variation reveals that each of the characters that in combination seem to distinguish these populations as separate taxa occurs in other populations as well. For example, *C. kucyniakii* is distinguished by its inflorescence of many flowers, smaller corollas with purple spots at the sinuses of the lobes, and large ovate-orbicular leaves. No other population of *C. strigosa* has all of these features in combination, or even to the extreme of the specimens representing *C. kucyniakii*; however, collections characterized by just one of the four traits that make *C. kucyniakii* distinct can be found throughout the range of the species. This implies that the genetic control for these traits is not unique to *C. kucyniakii* alone. Much more research is needed to understand the full genetic control and homology of these characters before a fully plausible explanation for the variation seen in this species can be proposed.

In the cladistic analyses of both morphology and cpDNA restriction site variation *C. kucyniakii* was treated as an independent taxonomic unit from *C. strigosa;* both analyses placed the two as sister taxa (Smith & Sytsma 1994a, b, c). Analysis of cpDNA variation placed *C. kucyniakii* as the sister taxon to an Ecuadorian accession of *C. strigosa* and these two accessions as the sister group to two Venezuelan accessions of *C. strigosa*. These results indicate that the *C. kucyniakii* variant is most closely related to, and derived from, the Ecuadorian populations of *C. strigosa* (Smith & Sytsma 1994a).

Phylogenetically, *C. strigosa* appears as an intermediate between sections *Pentadenia* and *Stygnanthe*. Based on cpDNA variation alone, *C. strigosa* appears as a clade within a large polytomy of other clades (Smith & Sytsma 1994b). Cladistic analysis of morphological variation places the *C. strigosa* clade more basally in the tree (Smith & Sytsma 1994a), and the combined analysis of morphology and cpDNA places the clade in a polytomy with the clade comprising *C. nervosa* and *C. isernii* and the clade containing sect. *Stygnanthe* (Fig. 3)(Smith & Sytsma 1994c).

9. Columnea trollii Mansfeld, Repert. Spec. Nov. Regni Veg. 36: 122. 1934. *Pentadenia trollii* (Mansfeld) Wiehler, Phytologia 27: 315. 1973.—Type: Bolivia. La Paz: Negracota, *Troll 1676* (holotype: B, destroyed; lectotype, here designated: M!).

Suffrutescent, sublignose, epiphytic or terrestrial herbs with upright ascending stems to 2 m tall, 4–9 mm in diameter, terete, tawny-brown, proximally smooth and glabrous, distally villous to appressed-pubescent with either red or transparent uniseriate trichomes; internodes 0.6–6.8 cm long; nodes swollen; leaf scars raised. Leaves opposite, isophyllous to slightly anisophyllous; laminas 4.3–13.4 cm long, 1.5–6.4 cm wide, elliptic, rarely lanceolate, rarely ovate, apex acute to acuminate, base rounded to cuneate, slightly oblique, adaxially green, appressed-hirsute, rarely strigose with red or transparent trichomes, abaxially pale green-maroon, strigose with single-celled transparent trichomes or hirsute with uniseriate transparent trichomes, veins abaxially villous to appressed-hirsute with red uniseriate trichomes, 6–7 lateral veins per side; margin serrulate to entire with red uniseriate trichomes; petioles 0.6–2.8 cm long, with red uniseriate trichomes. Inflorescences each of 1 flower in either axil of a leaf pair, rarely of 2 flowers; floral bract 1, 3–7 mm long, 0.6 mm wide, frequently caducous, linear, hirsute. Pedicels (0.6–) 2–5.8 cm long, erect in leaf axil, pubescent to villous, glands ca. 1 mm, fusiform, dark. Calyx clasping corolla; lobes equal to subequal, 1.1–2.7 cm long, 0.2–0.8 cm wide, lanceolate, acuminate, green, with red tips, or maroon, hirsute, margin minutely denticulate to entire. Corolla 2.7–5.3 cm long, 1.25–1.8 cm wide at widest point, 4–

7 mm wide before limb, 4–7 mm wide at base, tubular, ventricose, constricted at mouth and base, red-orange with yellow limb and lobes, exterior surface pilose with uniseriate transparent trichomes, more densely pilose at limb, interior slightly pilose at base, with capitate-glandular trichomes dorsally and distally; lobes equal, 1.5–3 mm long, 2–4 mm wide, semiorbicular. Filaments connate at base for 4 mm, adnate to base of corolla tube for 1 mm, white, glabrous; anthers 2 mm long, 2 mm wide, quadrate, included in corolla tube. Ovary 5–6 mm long, lanate; style white, glabrous to slightly pilose proximally, glandular distally; stigma bilobed or stomatomorphic, papillose, included in corolla tube. Nectary of 5 free glands, the 2 dorsal glands enlarged. Berry 0.8–1.2 cm in diameter, globose, pale green with purple stripes, pubescent; seeds 1.1–1.3 mm long, fusiform-oblong, twisted, red-brown, striate.

Phenology. Flowering from November to May, possibly earlier and later; fruiting December to May.

Distribution (Fig. 16). Bolivia (2450–3300 m) and Peru (1200–2000 m); cloud or elfin forests.

Additional Specimens Examined. **Peru.** Cuzco: Paucartambo, Pilla wata, Paso del Aguila, *Vargas C. 16801* (US); Prov. La Convención, 139 km from Cuzco, ascending toward the ridge between Santa Teresa and Chaullay, *Núñez & Motocanchi 8808* (MO, WIS). **Bolivia.** Cochabamba: Prov. Chaparé, Km 104 on rd into Chaparé from Cochabamba, *Steinbach 624* (F, MO, NY, U, US, WIS). Prov. Charasco: below Monte Puncu, along Río Lope Mendoza, *Besse et al. 657* (SEL); Siberia–Cochabamba, *Cardenas 5182* (US); 75.5 km E of Epizana on Carretera Fundamental 4, *Davidson 5113* (LAM, MO).—La Paz: Prov. Murillo: Valle de Zongo, Santa Rosa 3 km from La Paz, *Beck 1084* (SEL); N of La Paz, *Luer et al. 4957a* (SEL); 23.8 km N of pass from La Paz, *Solomon 16381* (MO, US); 24.3 N of pass from La Paz, *Solomon 17475* (MO, US). Prov. Sud Yungas: near Unduavi, *Besse et al. 621* (SEL); old rd to La Paz, SW of Coroico, *Luer et al. 5119* (SEL), *Luer et al. 5100* (SEL); rd from Cotopata to Coroico, *Smith & Beck 1723* (BH, LPB, US, WIS); rd from Cotopata to Chuspipata, *Smith & Smith 1828* (LPB, WIS); 0.9 km W of Chuspipata, *Solomon 9669* (MO, US, WIS); 3.4 km W of Chuspipata (1.5 km E Cotapata), *Solomon & Uehling 12276* (MO, US).

Columnea trollii may be distinguished from other species by its large, strongly inflated corollas and hirsute elliptic leaves. The two collections from Peru are slightly different from the Bolivian specimens in that they have the longest corollas seen of any specimen and are the only specimens with ovate leaves; however, these differences do not merit taxonomic recognition.

Columnea trollii is similar to *C. hypocyrtantha*, but the corolla of *C. trollii* is longer and the ventricose pouch is positioned more toward the middle of the corolla and not projecting forward, as it is in *C. hypocyrtantha.* In addition, *C. trollii* is characterized by hirsute pubescence but the ovary is lanate, whereas *C. hypocyrtantha* is nearly glabrous and the ovary entirely glabrous.

Cladistic analysis of morphology and cpDNA restriction site variation indicates that *C. trollii* is closely related to *C. oblongifolia* and is in a basally located clade with respect to all other species of *Columnea* (Fig. 2) (Smith & Sytsma 1994a, b, c). The overall morphology of this species clearly suggests its phylogenetic position as an intermediate between *Columnea* and *Alloplectus*. Many morphological features of *C. trollii* can be interpreted as a progression from *Alloplectus* to *Columnea*. The robust, shrubby herbaceous habit is similar to that of species of *Alloplectus* and the ventricose corolla is highly reminiscent of the urceolate corollas of many species of

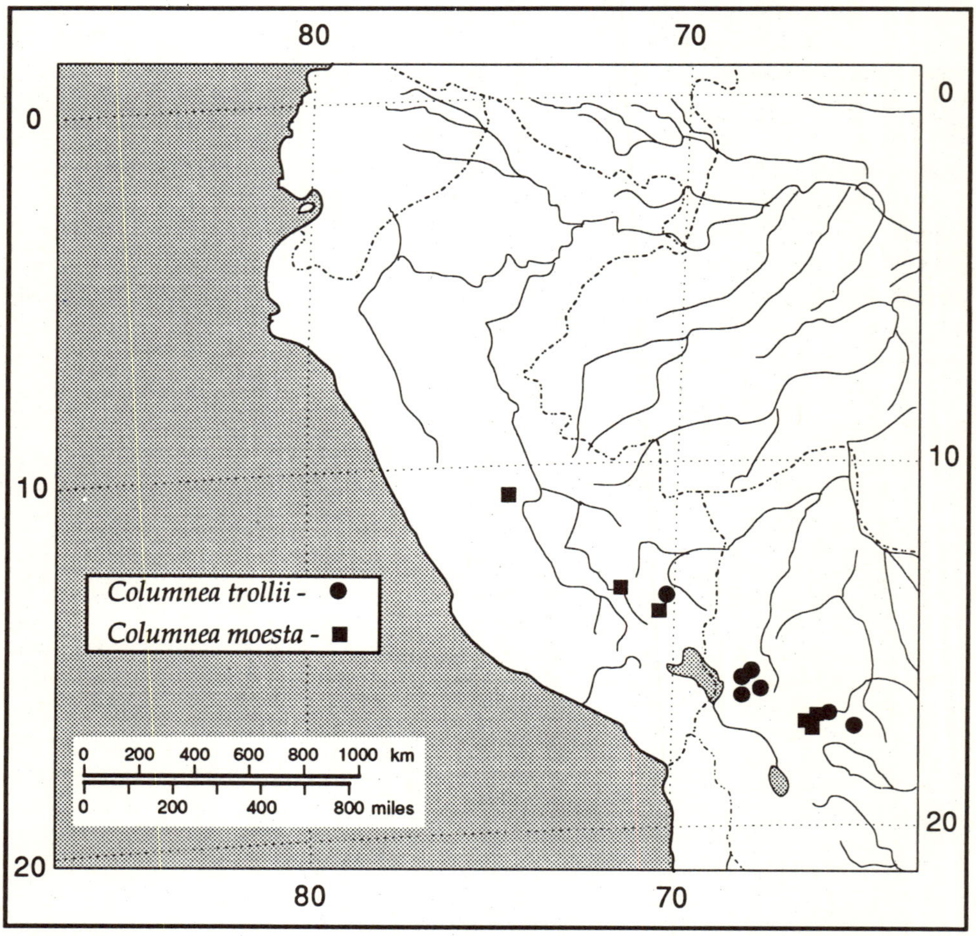

FIG. 16. Distribution of *Columnea trollii* and *C. moesta*.

Alloplectus. The urceolate corollas of *Alloplectus* are ventricose but not constricted, as in *C. trollii* and other species of *Columnea*.

The most striking intermediate feature between *Columnea* and *Alloplectus* is the fruit of *C. trollii*. The distinction between *Alloplectus* and *Columnea* is based on the dehiscence of the fruit. *Columnea* fruits are indehiscent berries, whereas the fruits of *Alloplectus* are fleshy and dehiscent (Wiehler 1983). Although the berries of *C. trollii* are indehiscent when fresh, they tend to split open along loculicidal lines that divide the fruit evenly when dried and pressed. Attempts to force the fresh mature berry to dehisce by squeezing were unsuccessful, and, instead, the berry became detached from the receptacle (pers. obs.); however, after drying these berries split open along well-defined lines in two specimens (*Steinbach 624*, *Smith & Smith 1828*). This character most likely represents a transition from the fleshy dehiscent display fruit of *Alloplectus* to the indehiscent berry of *Columnea*. Dehiscence itself has been lost, but the underlying mechanism for dehiscence remains in a vestigial state.

The placement of *C. trollii* in *Columnea* or *Alloplectus* becomes uncertain and arbitrary at this point. It should remain in *Columnea* until a thorough study of *Alloplectus* is completed.

Columnea section **Stygnanthe** (Hanstein) Bentham, Gen. pl. 2: 1010. 1876. *Stygnanthe* Hanstein, Linnaea 26: 185, 209. 1854.—TYPE: *Columnea moesta* Poeppig.

Occasionally sublignose (rarely succulent) epiphytes, occasionally epipetric or terrestrial herbs, rarely small shrubs, with creeping, upright, or pendent stems 0.4–3 m long, 1.5–15 mm in diameter, tan, green, red-brown, or dark purple, squarish-terete when dried, proximally smooth or flaking and glabrous (rarely hirsute), distally glabrescent to sericeous, with red or transparent uniseriate trichomes; internodes 0.4–9.5 cm long, rarely swollen. Leaves opposite or whorled, occasionally dorsiventrally arranged, anisophyllous or slightly anisophyllous; larger laminas 1.7–16.2 cm long, 0.8–6 cm wide, slightly falcate in some species, orbicular to elliptic, or oblanceolate to obovate, apex acute to acuminate or obtuse, base cuneate or rounded, oblique, adaxially light green suffused with rose, or green, glabrate to lanate with uniseriate red or transparent trichomes, or strigose with single-celled trichomes, abaxially green, purple, with purple apex, or reddish, glabrate to sericeous with red or transparent uniseriate trichomes, or strigose with single-celled trichomes, veins with red or transparent uniseriate trichomes usually longer than on lamina, lateral veins 3–10, margin entire, crenate, serrate, or undulate; smaller laminas 0.6–4.2 cm long, 0.25–2.4 cm wide, linear to ovate, apex acuminate, base rounded-oblique, otherwise similar to larger leaves; petioles absent or up to 31 mm long, hirsute to sericeous to lanate with transparent or red uniseriate trichomes. Inflorescences of 1–12 flowers in either axil of a leaf pair, commonly in the axil of the larger leaf; floral bracts 2–3, frequently caducous, 2.5–19 mm long, 0.3–9 mm wide, linear to ovate, apex acute, green or red, sparsely hirsute to sericeous with red or transparent uniseriate trichomes, margin entire. Pedicels absent or up to 28 mm long, erect, hirsute to lanate with uniseriate red or transparent trichomes, glands near calyx, 0.5–0.8 mm long, oval, purple, or eglandular. Calyx loosely clasping corolla; lobes equal to subequal, 6–21 mm long, 0.4–6 mm wide, linear to narrowly elliptic, spatulate or oblanceolate, apex acute or acuminate, green, purple, green with purple tips, or red, exterior surface hirsute to sericeous, interior glabrous to sparsely hirsute, or glandular, margin entire to serrate, or serrate-dissected only at base. Corolla 1.4–5.2 cm long, 2.5–15 mm wide at widest point, 2–7 mm wide before limb, 1–6 mm wide at base, tubular, only slightly ventricose, constricted at base, yellow, orange, red-purple, red-violet, lavender, orange-red, or red, limb frequently yellow-green, exterior surface sparsely hirsute to sericeous with red or transparent uniseriate trichomes, sometimes denser toward limb, interior slightly hirsute at base, glandular distally, or entirely glandular; lobes subequal, 1–3 mm long, 1–4.5 mm wide, semiorbicular, frequently with spot colored a darker shade than the corolla. Filaments connate at base for 0.7–10 mm, adnate to base of corolla tube for 1–5 mm, white, pink-purple, yellow, or red, glabrous to slightly pilose; anthers 0.6–2.1 mm long, 0.5–2.1 mm wide, quadrate, subquadrate, rectangular, included in corolla, rarely exserted. Ovary 1.5–3.5 mm long, glabrate to sericeous; style white, yellow, pink, or red, minutely pilose and glandular especially distally; stigma stomatomorphic or bilobed, white, yellow, red, or green, papillose, or smooth, included in corolla, rarely exserted. Nectary of 5 free glands (occasionally reduced to 2 dorsal glands), the 2 dorsal glands connate or free. Berry 5–17 mm long, 3–12 mm in diameter, globose or ovoid, pale lavender, violet, pink, lavender-blue, or white, glabrate to pilose; seeds 0.8–1.6 mm long, fusiform, oblong, twisted, brown-yellow or red-purple, striate.

The species in sect. *Stygnanthe*, as defined here, are many of those that Wiehler (1973, 1977, 1981, 1984) placed in the genus *Pentadenia* and Kvist and Skog (1993) transferred to sect. *Stygnanthe*. The circumscription of the section is similar to that of Kvist and Skog (1993), but excluding some species here assigned to sect. *Pentadenia* and including species not considered in their Ecuadorian treatment.

10. Columnea ambigua (Urban) Morley, Proc. Roy. Irish Acad. 74B(24): 423. 1974. *Alloplectus ambiguus* Urban, Symb. Antill. 1: 408. 1899. *Alloplectus ambiguus* var. *chlorosepalus* Urban, Symb. Antill. 1: 408. 1899, nom. superfl. *Crantzia ambigua* (Urban) Britton, Britton & Wilson, Sci. Surv. P. R. & V. I. 6: 204. 1925. *Ortholoma ambiguum* (Urban) Wiehler, Phytologia 27: 320. 1973. *Trichantha ambigua* (Urban) Wiehler, Selbyana 1(1): 34. 1975.—Type: Puerto Rico. *Eggers 1303* (lectotype, here designated: US!).

Alloplectus ambiguus var. *erythrosepalus* Urban, Symb. Antill. 1: 408. 1899.—Type: Puerto Rico. *Eggers 1302* (lectotype, here designated: US!).

Creeping, sublignose, epiphytic and terrestrial herbs with stems 3–8 mm in diameter, squarish when dried, proximally smooth and glabrous, distally hirsute with red or transparent uniseriate trichomes; internodes 0.8–8.8 cm long; nodes flush with stem; leaf scars raised. Leaves opposite, anisophyllous; larger laminas 3.5–11.5 cm long, 1.8–4.9 cm wide, oblong to elliptic, apex acuminate, base cuneate or rounded, oblique, adaxially green, sparsely hirsute to strigose with uniseriate red or transparent trichomes, abaxially green or reddish, sparsely hirsute to strigose with transparent uniseriate trichomes, veins with red uniseriate trichomes, lateral veins 4–6, margin crenate to serrate; smaller laminas 1.6–4.2 cm long, 0.85–2.4 cm wide, ovate to elliptic, apex acuminate, base rounded-oblique, otherwise similar to larger leaves; petioles 2–16 mm long, hirsute with transparent or red uniseriate trichomes. Inflorescences of 2–6 flowers in either axil of a leaf pair, commonly in the axil of the larger leaf; floral bract 1, 3–9 mm long, 1–4.5 mm wide, lanceolate to ovate, apex acute, green or red, sparsely hirsute with red uniseriate trichomes, margin entire. Pedicels 1–13 mm long, hirsute to sericeous with uniseriate red or transparent trichomes, eglandular. Calyx loosely clasping corolla; lobes unequal, 6–11 mm long, 1.5–3 mm wide, narrow elliptic to lanceolate, apex acute, green or red, exterior surface hirsute, interior glabrate, margin coarsely toothed. Corolla 1.4–2.2 cm long, 3–5 mm wide at widest point, 3–4 mm wide before limb, 1.5–2.5 mm wide at base, tubular, constricted at base, yellow, exterior surface sparsely hirsute with red or transparent uniseriate trichomes, interior slightly hirsute at base, glandular distally; lobes subequal, 1–2 mm long, semi-orbicular. Filaments connate at base for 3 mm at base, adnate to base of corolla tube for 2 mm, white, glabrate to slightly pilose; anthers 1.2 mm long, 1.5 mm wide, subquadrate, included in corolla tube. Ovary 1.5–2 mm long, pilose; style white, minutely pilose and glandular; stigma stomatomorphic, white, papillose. Nectary of 2 dorsal connate glands and also 1 or 2 free glands. Berry 7–8 mm in diameter, globose, white, glabrate; seeds not seen.

Phenology. Flowering from March to October.

Distribution (Fig. 17). Puerto Rico; 900–1050 m.

ADDITIONAL SPECIMENS EXAMINED. **Puerto Rico.** Monte Cerrote, near Adjuntas, *Britton & Brown 5415* (MO); Sierra Luquillo, 1 mi NW of Pico del Oeste, *Hartley 13321* (MO); Sierra Luquillo, *Howard 16815* (MO, WIS); El Yunque, *Otero 281* (MO); Sierra Luquillo, *Pfeifer et al. 2471, 2603, 2836, 2860, 3014* (CONN).

Columnea ambigua, endemic to Puerto Rico, is very similar to *C. angustata*, of Central and South America, but is readily distinguished by its coarsely toothed calyx lobes. Cladistic analysis of morphology (Smith & Sytsma 1994a) shows that *C. ambigua* is closely related to *C. angustata*.

11. Columnea angustata (Wiehler) L. E. Skog, Ann. Missouri Bot. Gard. 65: 85. 1979 ["1978"]. *Pentadenia angustata* Wiehler, Selbyana 2: 118. 1977.—TYPE: COLOMBIA. Valle del Cauca: 8 km past La Elsa, old rd from Cali to Buenaventura, *Wiehler et al. 7276* (holotype: SEL!).

Columnea sericea Mansfeld, Biblioth. Bot. 116: 145. 1937, non *Columnea sericea* (Hanstein) Kuntze, 1891. *Pentadenia sericea* (Mansfeld) Wiehler, Phytologia 27: 315. 1973.—TYPE: ECUADOR. Tungurahua: Río Negro, *Diels 878* (holotype: B, destroyed).—ECUADOR. Chimborazo: Naranjapata, Río Chanchan, 1933, *Schimpff 523* (neotype, designated by Kvist & Skog, 1993: M!; isoneotypes: MO, TRT, GH!).

Pentadenia ecuadorana Wiehler, Selbyana 2: 82. 1977. *Columnea ecuadorana* (Wiehler) L. E. Skog, Taxon 33: 126. 1984.—TYPE: ECUADOR. Pastaza: Puyo, *Wiehler et al. 7163* (holotype: SEL!).

Small, shrubby, epiphytic, rarely terrestrial herbs with ascending or rarely pendent stems to 2.5 m long, 2–12 mm in diameter, red-brown, squarish when dried, proximal epidermis smooth to flaking, glabrous, rarely hirsute, distally pubescent to hirsute with uniseriate red trichomes; internodes 0.4–7.7 cm long; nodes flush with stem; leaf scars raised. Leaves opposite, slightly anisophyllous; laminas 1.7–13.2 cm long, 0.8–3.6 cm wide, elliptic, rarely oblanceolate, occasionally ovate, apex acute to acuminate, base cuneate, rarely rounded, oblique, adaxially green, glabrate to sparsely hirsute, rarely strigose with uniseriate transparent trichomes, abaxially sericeous, rarely pilose with uniseriate red or transparent trichomes, lateral veins 3–5, margin entire; petioles 0.2–3.1 cm long, sericeous to hirsute with uniseriate red or transparent trichomes. Inflorescences of 1–4 (–6) flowers in either axil or in both axils of a leaf pair; floral bracts 2–3, 2.5–15 mm long, 0.5–6.5 mm wide, ovate to lanceolate, apex acute, red or green, sericeous with uniseriate transparent trichomes, margin entire. Pedicels 1–10 mm long, sericeous with uniseriate transparent trichomes, eglandular. Calyx loosely clasping corolla; lobes equal to subequal, 6–18 mm long, 1–6 mm wide, lanceolate, apex acuminate, red, rarely green, exterior surface sericeous, occasionally less densely sericeous toward apex, interior glabrate to sparsely hirsute, margin serrate at base of lobe. Corolla 1.9–3.1 cm long, 2.5–6 mm wide at the widest point, 2–4.5 mm wide before limb, 1–3 mm wide at base, tubular, constricted at base, widening at mouth, yellow, orange, or red, exterior surface sericeous with uniseriate red or rarely transparent trichomes, interior with ring of trichomes at base and with glandular trichomes over the remainder of the surface; lobes subequal, 1–3 mm long, 1–4 mm wide, semiorbicular. Filaments connate at base for 2–4 mm, adnate to base of corolla tube for 1–3 mm, white or red, sparsely pilose; anthers 0.7–1.5 mm long, 0.7–1.5 mm wide, quadrate, included in corolla tube. Ovary 2.5–3 mm long, sericeous with uniseriate red or transparent trichomes; style white or red, proximally sparsely pilose, distally glandular; stigma bilobed, green, smooth, included in corolla tube. Nectary variable, of 5 free glands or 2 dorsal connate and 2 free glands. Berry 5–12 mm long, 3–12 mm in diameter, globose to ovoid, pale lavender, pilose to glabrate; seeds ca. 1.1 mm long, fusiform to oblong, brown-yellow or red-purple, striate.

Phenology. Flowering from March to October in Central America, December to June (one collection in August) in Colombia, and continuously in Ecuador.

Distribution (Fig. 17). Costa Rica to Ecuador; wet montane forests; 0–1800 m.

REPRESENTATIVE SPECIMENS. **Costa Rica.** CARTAGO: along Río Pejibaye between Río Taus and Quebrada Azul, *Lent 2551* (SEL).—HEREDIA: Cariblanco, valley of Río Sarapiqui, cultivated at BH, *Moore 7773* (BH, US). **Panama.** COCLÉ: near Asserradero El Copé, 8 km N of El Copé, *Dressler 5642* (SEL, US).—DARIÉN: N slopes of Cerro Pirre, *Mori & Kallunki 5471* (MO, US).—VERAGUAS: 15–20 km NW of Santa Fé, between Escuela Agrícola Alto Piedra and continental divide, *Dressler 4732* (SEL); 5 mi W of Santa Fé on rd past Escuela Agrícola, Pacific side of divide, *Liesner 916* (MO). **Colombia.** ANTIOQUIA: Mutatá, 1 km rd from Mutatá to Pavarandó, *Fonnegra et al. 1865* (MO, US, WIS); San Luis, *Orozco et al. 618* (COL).—CALDAS: Tatamá, Sta. Cecilia, *Sneidern 4991* (S).—CAUCA: Río Timbiquí, *Lehmann s.n.* (K).—CHOCÓ: Quibdó, Río Atrato, *Archer 1904* (US); Mpio. Tadó, 27 km E, *Bernal et al. 1034* (COL); Río San Juan, small quebrada across from Palestina, *Cuatrecasas 21359* (F, US); left bank of Río Atrato, across from Quibdó, Barrio "Avenida Bahía Solano," *Forero et al. 1459* (COL, MO, US); rd San Jose del Palmar to Nóvita, Río Ingará, between La Italia and Curundó, *Forero et al. 2111* (COL, MO); Quebrada La Sierpe, across from Palestina, *Forero et al. 3971* (COL); confluence of ríos Tamaná and San Juan, between Primavera and Sta. Rosa, *Forero et al. 4903* (COL, MO); valley of Río San Juan, Andagoya, mining camp la Cia, *Forero et al. 5126* (COL, MO); Bojayá, Quebrada del Puerte up from Bellavista, *Forero et al. 9295* (COL); right bank of Río Baudó, 18.5 km upstream of estuary, between Quebrada and sawmill of Porquera, *Fuchs & Zanella 21792* (COL, US); San José del Palmar, Tereda Torito, finca la Guaduales, *Guerra 37* (COL).—NARIÑO: Reserva La Planada, Km 63 on Tumaco–Túquerres rd, *Gentry et al. 34920* (COL).—VALLE DEL CAUCA: Río Naya upriver from Puerto Merizalda, *Gentry & Juncosa 40690* (COL, MO, US); hwy Cali to Buenaventura, *Killip & Cuatrecasas 38996* (US); Buenaventura, San Isidro, near INDERENA-FAO camp, *Rooden et al. 449* (COL, US); area controlled by Corporación Valle del Cauca (El Chanco), trail along Río Calima from Campoalegre to Campamente Cuzumbu, across river, *Smith et al. 1433* (AAU, BH, COL, SEL, TULV, US, WIS). **Ecuador.** BOLÍVAR: Charquiyacu, *Acosta S. 6095* (F).—COTOPAXI: Río Guapara, ca. 20 km NW of El Corazón, *Sparre 17099* (S).—EL ORO: 11 km W of Piñas, on new rd to Sta. Rosa, *Dodson et al. 9067* (MO, SEL).—LOS RÍOS: Río Palenque Biological Station, Km 56 Quevedo–Santo Domingo, *Dodson 5667* (SEL); near Los Angeles, Km 51 Quevedo–Santo Domingo, 4 km W of hwy, *Gentry & Dodson 18026* (MO); near Montalvo, 40 km E of Babahoyo, *Holm-Nielsen et al. 2679* (F, MO, S, US).—MANABÍ: slopes of Montecristi, *Besse et al. 1274* (SEL).—MORONA-SANTIAGO: Cordillera de Cutucú, trail from Logroño to Yaupi, *Madison et al. 3192* (SEL).—PASTAZA: Mera, *Asplund 18932* (K, NY, S); Cajabamba-Mariscal rd, departing main Tena–Puyo rd at km 31 N of Puyo, ca. 2 km from main hwy, *Croat 58926* (MO); Mera, trail to Pindo, *Harling et al. 19617* (GB); Veracruz (Indillama–)Canelos, 8 km E of Veracruz, *Lugo S. 1111* (MO, SEL); Moravia, rd Mera–Puyo, 5 km SE of Mera, *Lugo S. 1805* (GB, SEL); Canelos, *Lugo S. 4506* (SEL); Motolo, near Shell–Mera, *Lugo S. 676* (SEL, WIS).—PICHINCHA: 2 km SE Santo Domingo, Cooperativa Santa Marta #2, along Río Verde, *Dodson 7407* (MO, SEL); 5 km S of Santo Domingo, Hacienda San Fernando, *Hansen et al. 7851* (MO, SEL); Finca

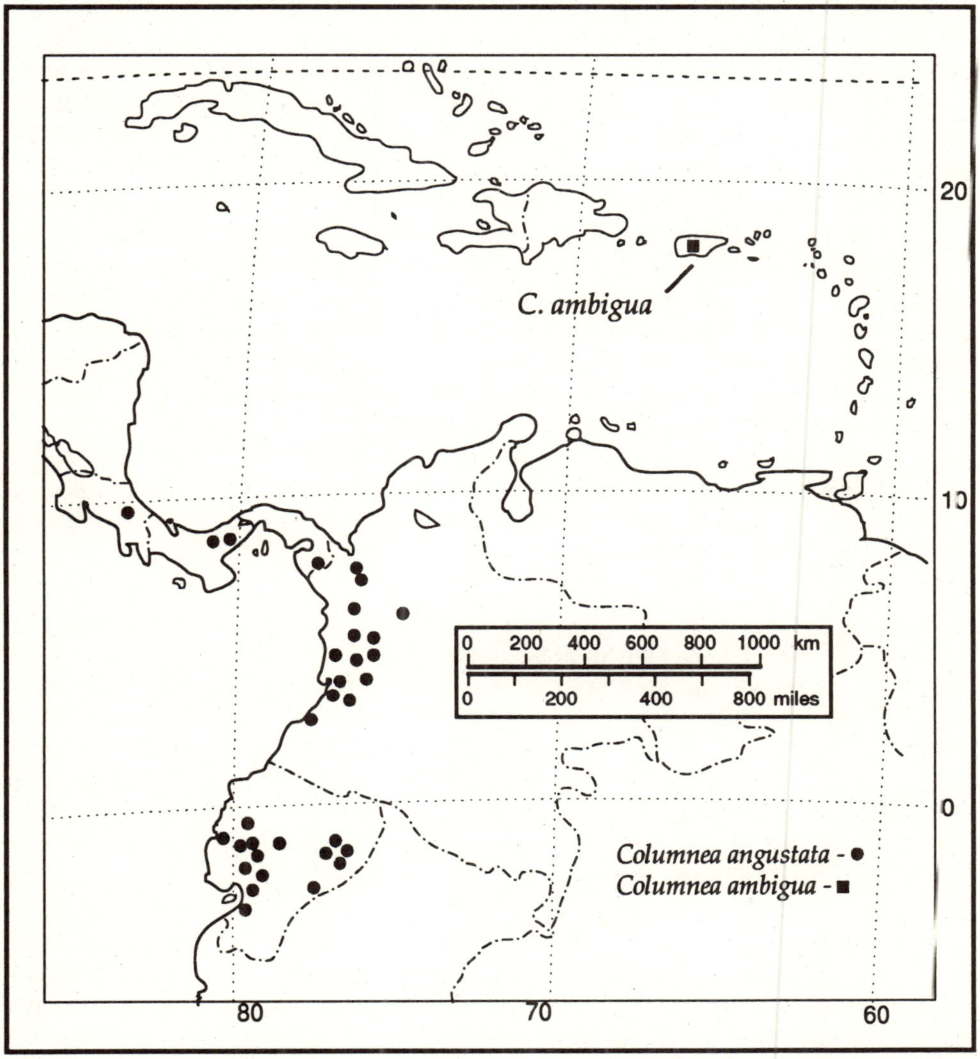

FIG. 17. Distribution of *Columnea angustata* and *C. ambigua*.

La Carlita, 11 km W of Santo Domingo on rd to Chone, *Iltis & Iltis E–252c* (SEL, WIS); Santo Domingo, Hacienda Zaracay, *Sparre 15144* (S).—TUNGURAHUA: Río Negro, *Asplund 8554f* (S, US); near Río Margarjitas, along Canelos trail, *Penland & Summers 145* (F, US).

Columnea angustata is a morphologically variable species but readily distinguished from other species of *Columnea*. The relatively small, yellow, orange, or red corollas in combination with leaves that are slightly but not strongly anisophyllous distinguish *C. angustata*. It is most similar to the Puerto Rican endemic *C. ambigua*, which differs in its coarsely toothed calyx lobes.

Columnea angustata is similar to *C. spathulata* in that it is a widespread species with several morphological variants. These variants have numerous intermediates that form a continuum; thus, recognition of the variants as species is extremely difficult and impractical. In addition, analysis of cpDNA restriction site variation of representatives of three morphological variants ranging from Costa Rica to Ecuador indicates a close relationship among these taxa. All specimens of *C. angustata* used in the analysis were identical with the exception of one accession, W2479, which lacked a mutation shared by the other three (Smith & Sytsma 1994b).

Cladistic analysis of cpDNA variation indicates that *C. angustata* is the sister species to *C. spathulata* (Smith & Sytsma 1994b). These species share several morphological features, such as thickened square stems and globose white berries. Yet, in a cladistic analysis of morphology, *C. angustata* was the sister species to the remaining species of sect. *Stygnanthe* (Smith & Sytsma 1994a). Because so few characters supported this placement, and many more cpDNA characters supported its status as a sister species to *C. spathulata*, the combined analysis favored the cpDNA position (Fig. 2) (Smith & Sytsma 1994c).

12. Columnea antiocana (Wiehler) J. F. Smith, comb. nov. *Pentadenia antiocana* Wiehler, Selbyana 7: 335. pl. 2D. 1984.—TYPE: COLOMBIA. Antioquia: *Jewise s.n.* (holotype: K!)

Vining herbs, stems 2–2.5 mm in diameter, terete, tan, proximally smooth and glabrous, distally glabrescent to sericeous, with numerous adventitious roots; internodes 1–5.3 cm long; nodes flush with stem; leaf scars raised. Leaves opposite, isophyllous to slightly anisophyllous; laminas 2.5–5.5 cm long, 0.8–1.6 cm wide, ovate to lanceolate or elliptic, apex acute to long-acuminate, base rounded, oblique or not, adaxially green, slightly hirsute to glabrous, abaxially green, strigose to glabrate, veins slightly hirsute to sericeous, lateral veins 3–4, margin entire; petioles 2–4 mm long, sparsely hirsute to sericeous. Inflorescences of 1 flower per axil; floral bract 1, 2.5–3 mm long, caducous, linear. Pedicels 4.5–20 mm long, erect, hirsute to sericeous, glands near calyx 0.5–0.7 mm long, oval, purple. Calyx loosely clasping corolla; lobes equal to subequal, 9–13 mm long, 2–2.5 mm wide, lanceolate, apex acute, green, exterior surface strigose to hirsute, interior glabrous, margin entire. Corolla 2.5–3 cm long, 3–6 mm wide at widest point, 3–4 mm wide before limb, 1.5–2 mm wide at base, tubular, constricted at base, slightly constricted before limb, red, exterior surface sparsely pilose to slightly sericeous with red uniseriate trichomes, interior glabrous; lobes equal to subequal, 1.5–2 mm long, 1.5–2 mm wide, semiorbicular, with darker red spots on interior surface. Filaments connate at base for 4 mm, adnate to base of corolla tube for 1 mm, white, glabrous; anthers 0.6–1.5 mm long, 0.6–1.5 mm wide, quadrate, included in corolla tube. Ovary 3 mm long, sericeous; style white, glabrous; stigma stomatomorphic, white, smooth, included in corolla tube. Nectary a single dorsal connate gland. Fruit not seen.

Distribution (Fig. 18). Colombia (Antioquia, Valle del Cauca); ca. 2000 m.

ADDITIONAL SPECIMENS EXAMINED. **Colombia.** VALLE DEL CAUCA: Hacienda Tokio, behind microwave tower, 10 km S of Queremal, *Gentry et al. 40833* (COL, MO); San Julian, *Triana 2477* (P).

The thin, probably pendent stems of *C. antiocana* are similar to those of *C. colombiana*, but *C. antiocana* may be distinguished by the entire margins of its calyx lobes and its more ovate, acute to acuminate laminas. Although it is superficially similar to *C. colombiana*, *C. antiocana* is more likely related to the group of species including *C. lavandulacea*, *C. ovatifolia*, and *C. crassicaulis* (Fig. 2), with which it shares a similar laminar shape, vesture, and the presence of darker colored spots on the lobes of the corollas.

13. Columnea byrsina (Wiehler) L. P. Kvist & L. E. Skog, Allertonia 6: 384. 1993. *Pentadenia byrsina* Wiehler, Selbyana 2: 119. 1977.—TYPE: [ECUADOR.] Cultivated material grown from seeds collected near Baeza, Napo, *Wiehler 77122* (holotype: SEL!).

Suffrutescent, epiphytic, occasionally terrestrial herbs with ascending and spreading, branching stems to 2 m long, 2–8 mm in diameter, with zigzag appearance, terete, maroon, proximally smooth and glabrous, distally sericeous with red uniseriate trichomes; internodes 0.7–9.5 cm long; nodes and leaf scars flush with stem. Leaves opposite, dorsiventrally arranged, strongly anisophyllous; larger laminas 4–8 cm long, 2.5–2.8 cm wide, oblanceolate to elliptic, apex acute to acuminate, base cuneate or rounded, oblique, adaxially dark green, slightly pilose to glabrous, abaxially lighter green suffused with rose-red, long-sericeous and short-tomentose vesture of red or transparent uniseriate trichomes, vesture appressed and somewhat denser on veins, lateral veins 6–10, margin entire, ciliate with red uniseriate trichomes; smaller laminas 1–2.8 cm long, 0.45–0.7 cm wide, linear to lanceolate, frequently absent, otherwise like the larger laminas; petioles absent to 3 mm long, sericeous with red or transparent uniseriate trichomes. Inflorescences of 2–8 flowers per axil; floral bract 1, up to 3 mm long, 0.5 mm wide, linear. Pedicels 6–13 mm long, erect, sericeous, rarely with oval purple glands near calyx. Calyx clasping corolla; lobes equal to subequal, 6–14 mm long, 0.7–1.5 mm wide, linear to lanceolate, acute, green, or flushed with pink, or purple, exterior surface sericeous, interior glabrous, margin entire. Corolla 1.8–2.9 cm long, 3–7 mm wide at widest point, 2.5–4 mm wide before limb, 1–2 mm wide at base, tubular, constricted at base and slightly before limb, bright red with lemon-yellow limb and lobes, exterior surface sericeous to hirsute with some glandular trichomes, interior with glandular trichomes dorsally and distally; lobes equal to subequal, 1–2 mm long, 1–2 mm wide, semiorbicular. Filaments connate at base for 2.5 mm, adnate to base of corolla tube for 1 mm, pilose from base to midpoint, then glabrous; anthers 0.7–2 mm long, 0.5–1.5 mm wide, rectangular, exserted 5–9 mm beyond opening of corolla. Ovary 1.5–2 mm long, sericeous to glabrate; style white, pilose; stigma bilobed, smooth, exserted 5–7 mm beyond opening of corolla. Nectary of 5 free glands, the 2 dorsal glands connate. Berry 8–10 mm long, 4–10 mm in diameter, ovoid or globose, white, slightly pilose to pubescent; seeds 1.6 mm long, oblong, twisted, light brown, striate.

Phenology. Flowering continuously.

Distribution (Fig. 18). Central Colombia to Ecuador; wet forests; 650–1950 m.

REPRESENTATIVE SPECIMENS. **Colombia**. ANTIOQUIA: Mpio. Frontino, Km 17 rd Nutibara to Murri, *Zarucchi et al. 5736* (MO, WIS).—CHOCÓ: rd Tutunendo to El Carmen, between Kms 135 and 120, *Forero et al. 6107* (COL).—NARIÑO: Reserva Natural La Planada, Ricaurte, *Benavides 8752* (US); Reserva Natural La Planada, Ricaurte, *Smith & Galeano 1505* (AAU, B, BH, COL, F, K, La Planada,

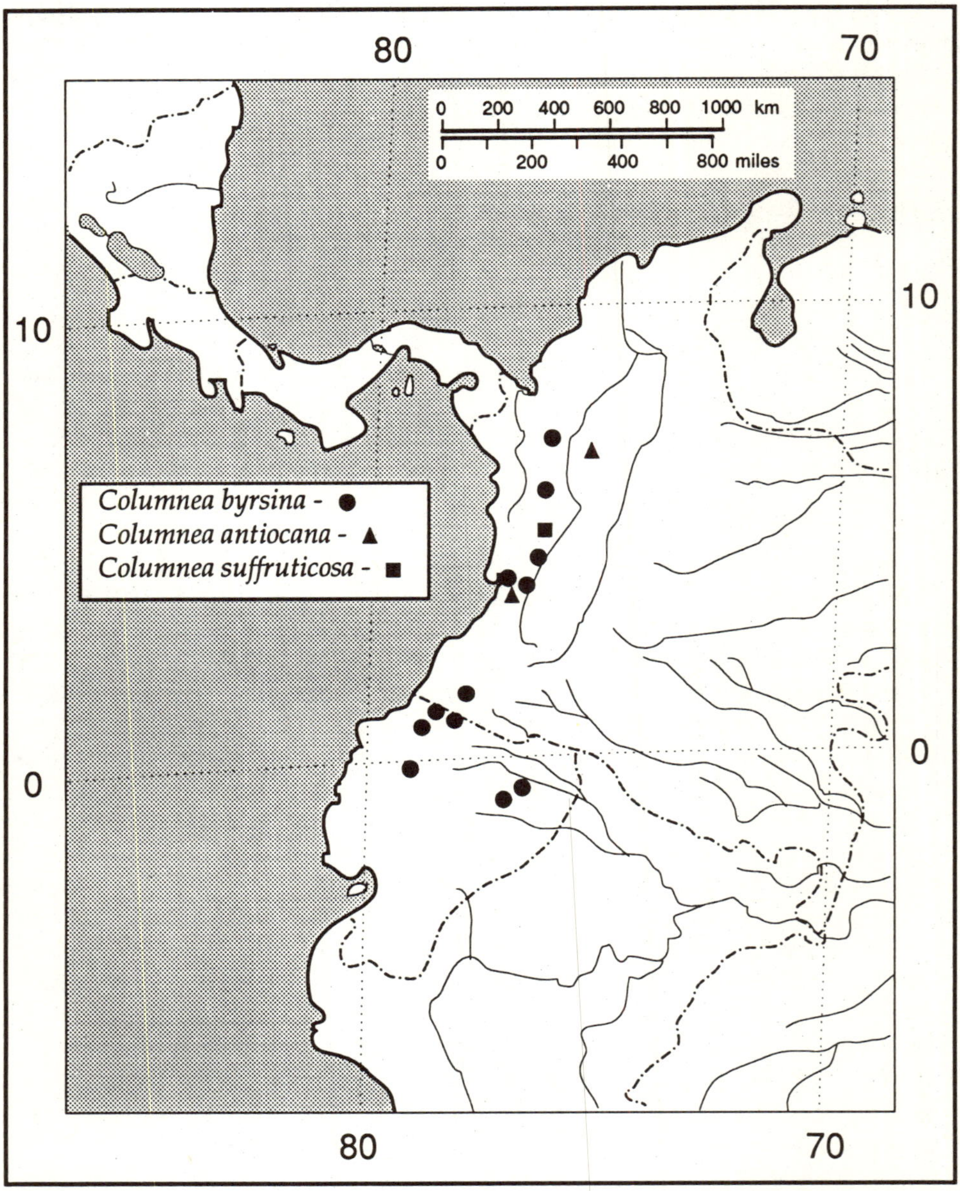

FIG. 18. Distribution of *Columnea byrsina, C. antiocana,* and *C. suffruticosa.*

MO, NY, PSO, QCA, S, SEL, TULV, US, WIS), *1519* (COL, La Planada, PSO, WIS).—VALLE DEL CAUCA: between Las Brisas and La Marina, *Cuatrecasas 22667* (F); valley of Río Dígua, left bank of Río San Juan near Queremal, *Cuatrecasas 23702* (F); Las Delicias, NW Restrepo, *Robinson 209* (COL, US). **Ecuador.** CARCHI: 90 km W Tulcán, near Maldonado, *Balslev 1997* (US); El Pailon, 45 km Maldonado, *Madison & Besse 7115* (SEL).—ESMERALDAS: Km 9–12, rd Lita to Alto Tambo, *Dodson et al. 16836* (MO).—NAPO: Puente Sardinas Grandes, Km 18 Baeza to Lago Agrio, *Balslev & Madsen* (COL, F, MO, SEL, U); Km 6 Baeza to Lago Agrio rd, *Besse et al. 1633* (SEL, US).—PICHINCHA: Chiriboga, old rd Quito to Santo Domingo, *Sauleda et al. 4011* (SEL).

Columnea byrsina is readily distinguished from other species by its anisophyllous, acute to acuminate leaves and bright red corollas with exserted stamens and styles. Flowers are located beneath the larger leaf of a pair. In full sun the flowers are in the open; in low light or shade they are covered by the larger leaf. This species is most closely related to *C. orientandina*, which also exhibits the corolla movements described above (Smith & Sytsma 1994a, b, c).

Columnea byrsina is the only species of *Columnea* known to have a variable fruit shape. The berries are generally globose, but several collections from near Baeza, Napo, Ecuador, have ovoid berries.

14. Columnea colombiana (Wiehler) L. P. Kvist & L. E. Skog, Allertonia 6: 385. 1993. *Pentadenia colombiana* Wiehler, Selbyana 2: 120. 1977.—TYPE: COLOMBIA. Valle del Cauca: along Río Dagua, old rd Cali–Buenaventura, near Buenaventura, 1 May 1972, *Wiehler et al. 72130* (holotype: SEL!).

Epiphytic herbs with vining or pendent stems to 3 m long, 1.5–3.5 mm in diameter, terete, green, with many adventitious roots, hirsute; internodes 0.4–3.2 cm long; nodes and leaf scars flush with stem. Leaves opposite, slightly anisophyllous; laminas 1.3–4.1 cm long, 0.8–2.4 cm wide, ovate to orbicular, apex acute, base rounded, oblique, adaxially and abaxially green, hirsute, lateral veins 3–4, margin entire; petioles 3–10 mm long, hirsute. Inflorescences of 1–4 flowers per axil; floral bract 1, 3–4.5 mm long, linear, green, hirsute. Pedicels 6–13 mm long, erect, green, hirsute, eglandular. Calyx loosely clasping corolla, the tips slightly recurved; lobes subequal, 6–14 mm long, 1–2.5 mm wide (excluding teeth), lanceolate to ovate, long-acuminate, green, tips and teeth red, exterior surface hirsute, interior glabrous, margin serrate and dissected at base. Corolla 2.3–3.2 cm long, 4–6 mm wide at widest point, 3–4.5 mm wide before limb, 1–2 mm wide at base, tubular, constricted at base, slightly constricted before limb, red-purple, limb and lobes green-yellow, exterior surface long-hirsute with uniseriate red trichomes and some glandular trichomes, trichomes on limb with red base, interior glabrous with glands distally and dorsally; lobes subequal, 1–1.5 mm long, 1.5–2 mm wide, semiorbicular. Filaments connate at base for 7 mm, adnate to base of corolla tube for 5 mm, yellow, glabrous to slightly pubescent; anthers 1 mm long, 1 mm wide, quadrate, included in corolla tube. Ovary 2–3 mm long, sericeous; style white, distally pilose to glandular; stigma bilobed, green, papillose, included in corolla tube. Nectary of 5 free glands, the 2 dorsal glands connate. Berry 1.3 cm in diameter, globose,white, pilose; seeds 1.1 mm long, oblong, twisted, brown, striate.

Phenology. Flowering from May to August, possibly longer.

Distribution (Fig. 5). Colombia and northern Ecuador; wet forests; sea level to 40 m.

ADDITIONAL SPECIMENS EXAMINED. **Colombia.** CHOCÓ: Boca Pepe, downstream of Porto Malua, Río Baudo, *White & Warner 784* (COL, MO); cultivated material, provenance unknown, *Smith 1126* (WIS).

The slightly anisophyllous, ovate to orbicular leaves, thin pendent stems, red corollas with green limb, and dissected calyx lobes distinguish *C. colombiana* from other species. Although morphologically it is distinct from other species of sect. *Stygnanthe*, and a cladistic analysis of morphology allies it with *C. rileyi* and *C.*

suffruticosa (Smith & Sytsma 1994a), the cpDNA analysis clearly places it as a sister taxon to *C. byrsina* and *C. orientandina* (Fig. 2) (Smith & Sytsma 1994b, c).

15. Columnea crassicaulis (Wiehler) L. P. Kvist & L. E. Skog, Allertonia 6: 385. 1993. *Pentadenia crassicaulis* Wiehler, Selbyana 2: 122. 1977.—TYPE: COLOMBIA. Nariño: *Wiehler & Williams 72185* (holotype: SEL!).

Succulent, epiphytic or epipetric herbs with creeping stems to 3 m long, 4–6 mm in diameter, terete, green or suffused with purple, proximally glabrescent, distally sparsely sericeous; internodes 0.5–3.5 cm long, swollen; nodes flush with stem; leaf scars raised. Leaves opposite, isophyllous to slightly anisophyllous; laminas 3.5–6.2 cm long, 1.8–4.4 cm wide, ovate to orbicular, apex acuminate to acute, base rounded, oblique, adaxially yellow-green, glabrous to strigillose, abaxially reddish, strigose to glabrate, veins strigose, lateral veins 3–6, margin subentire to crenulate; petioles 3–15 mm long, slightly hirsute to slightly sericeous. Inflorescences of a solitary flower per leaf axil, bract 1, 3–12 mm long, 0.5–1.5 mm wide, linear, apex acute, pink-red, margin entire. Pedicels 4–9 mm long, erect, sericeous, eglandular. Calyx clasping corolla; lobes equal to subequal, 5–18 mm long, 2–4.5 mm wide, lanceolate, apex acute, green-pink, exterior surface slightly pilose, interior glabrous, margin entire. Corolla 3.2–5.2 cm long, 4.5–10 mm wide at widest point, 4–6 mm wide before limb, 2–3 mm wide at base, tubular, ventricose, long constriction at base, slightly constricted before limb, lemon-yellow, limbs green, exterior surface appressed-pilose to sericeous, interior glabrous with glandular trichomes distally and dorsally; lobes equal to subequal, 1.5–2.5 mm long, 2.5–3 mm wide, semiorbicular with dark orange-yellow spots on interior surface. Filaments connate at base for 9–10 mm, adnate to base of corolla tube for 4–5 mm, slightly pubescent; anthers 1 mm long, 1 mm wide, quadrate, included in corolla tube. Ovary 2.5–3.5 mm long, glabrous; style white, glabrous; stigma stomatomorphic, red, smooth. Nectary of 5 free glands with the 2 dorsal glands connate. Berry 7–10 mm long, 5–7 mm in diameter, ovoid, white, glabrous; seeds 1.2 mm long, oblong, twisted, brown, striate.

Phenology. Flowering continuously.

Distribution (Fig. 5). Southern Colombia to northern Ecuador; wet forests; 1500–2000 m.

ADDITIONAL SPECIMENS EXAMINED. **Ecuador.** CARCHI: Maldonado, *Harling & Andersson 12391* (SEL).—PICHINCHA: near Mindo, *André 3526* (K); between Nanegalito and Gualea, *Dodson et al. 6983* (SEL); rd Nono to Tandayapa, *Dressler 4926* (SEL); below Nono on route to Mindo, *Luer et al. 4717* (SEL).

Columnea crassicaulis is distinguished by its thickened stem, ovate to orbicular leaves, and long yellow corollas with a long constriction at the base. Cladistic analysis of cpDNA variation indicates it is closely related to other Ecuadorian species, such as *C. lavandulacea*, *C. rileyi*, and *C. ovatifolia* (Fig. 2) (Smith & Sytsma 1994b).

16. Columnea fritschii (Rusby) J. F. Smith, comb. nov. *Fluckigeria fritschii* Rusby, Bull. Torrey Bot. Club 24: 488. pl. 126. 1894. *Kohlerianthus fritschii* (Rusby) Fritsch in Engler & Prantl, Nat. Pflanzenfam. Nachtr. 1: 300. 1897. *Pentadenia fritschii* (Rusby) Wiehler, Phytologia 27: 314. 1973.—

Type: Bolivia. La Paz: Yungas, *Bang 401* (holotype: NY; isotypes: F! GH! K! MO! US! W! WIS!).

Erect, elongated herbs with vining stems to 3 m long, 3–6 mm in diameter, terete, red-brown, proximally slightly pilose to tomentose, distally pilose to tomentose; internodes 2.0–6.8 cm long; nodes flush with stem; leaf scars raised. Leaves in whorls of 4, isophyllous to slightly anisophyllous; laminas 2.2–5.0 cm long, 1.2–2.7 cm wide, ovate, apex acute to acuminate, base rounded, slightly oblique, adaxially green, appressed hirsute on midrib, slightly glabrous to slightly hirsute to pilose, abaxially green, strigose to appressed-pilose, denser on veins, lateral veins 4–5 per side, margin entire, ciliate with red uniseriate trichomes; petioles 2–9 mm long, appressed hirsute. Inflorescences of a solitary flower per axil; floral bract 1, up to 5 mm long, 2 mm wide, caducous, lanceolate, green, sericeous. Pedicels 6–8 mm long, erect, sericeous with oval, purple glands near calyx. Calyx loosely clasping corolla, the tips recurved; lobes equal to subequal, 7–13 mm long, 3–4 mm wide, ovate to lanceolate, acuminate, green with purple tips, exterior surface densely hirsute, interior glabrous, margin entire to serrulate. Corolla 3.5–4.7 cm long, 10–12 mm wide at widest point, 6–7 mm wide before limb, 2.5–4 mm wide at base, tubular, ventricose, constricted at base and limb, limb not as wide as ventricose part of corolla tube, bright purple-red, limb yellow, exterior surface densely long-hirsute with red uniseriate trichomes, denser toward limb, interior slightly hirsute with glandular trichomes dorsally and distally; lobes equal to subequal, 2–3 mm long, 2–3 mm wide, semi-orbicular with purple spots on interior surface. Filaments connate at base for 4 mm, adnate to base of corolla tube for 3 mm, yellow, slightly pilose; anthers 1.7–2.1 mm long, 1.7–2.1 mm wide, quadrate, included in corolla tube. Ovary 2–2.5 mm long, sericeous; style yellow, hirsute at base but becoming glabrous, with capitate-glandular trichomes distally; stigma bilobed, smooth, included in corolla tube. Nectary of 5 free glands. Fruit and seeds not seen.

Phenology. Flowering October to November, possibly longer.

Distribution (Fig. 12). Bolivia (La Paz, Prov. Nor Yungas); cloud forests; 2000 m.

Additional Specimens Examined. **Bolivia.** La Paz: Prov. Nor Yungas, Polo-Polo above Coroico, *Buchtien 3902* (US); 16.2 km NE Chuspipata on rd to Yolosa, *Solomon 12531* (LPB, MO).

This species and *C. ultraviolacea* are the only two species of *Columnea* with leaves arranged in whorls of four. *Columnea fritschii* has bright red corollas and green leaves, whereas *C. ultraviolacea* has yellow corollas, purple leaves, and more lanceolate calyx lobes.

Both *C. fritschii* and *C. ultraviolacea* are known only from small areas in Bolivia and, because of their unusual leaf arrangement, are undoubtedly sister species, a suggestion supported by a cladistic analysis of morphology (Smith & Sytsma 1994a).

17. Columnea inconspicua L. P. Kvist & L. E. Skog, Allertonia 6: 385. 1993.— Type: Ecuador. Pichincha: Tandápi, confluence between Río Tandápi with Río Pilatón, 1500 m, 27 Jul 1967, *Sparre 17761* (holotype: S!).

Epiphytic herbs with frequently branching stems 4–6 mm in diameter, terete, red-brown, proximally smooth and glabrous, distally glabrate to pubescent; internodes 0.6–3.5 cm long; nodes flush with stem; leaf scars raised. Leaves opposite, strongly anisophyllous, occasionally slightly anisophyllous; larger laminas 2.7–8.3 cm long, 0.8–2 cm wide, lanceolate to slightly falcate, apex acuminate, base cuneate to rounded, strongly oblique, adaxially green, glabrous to sparsely pilose with uniseriate transparent trichomes, abaxially green, or occasionally with a red apex, strigose with single-celled transparent trichomes, veins sericeous with uniseriate transparent trichomes, lateral veins 3–6, margin undulate, ciliate with red-purple uniseriate trichomes; smaller laminas 8–16 mm long, 4–6 mm wide, otherwise like the larger laminas; petioles 0.5–5 mm long. Inflorescences of 1–5 flowers in axil of larger leaf of a pair; floral bract 1, 5 mm long, 1.2 mm wide, caducous, lanceolate, apex acuminate, green, villous to sericeous, margin entire. Pedicels 2–6 mm long, sericeous with uniseriate transparent trichomes, eglandular. Calyx loosely clasping corolla; lobes equal, 7–15 mm long, 1–1.5 mm wide, lanceolate to subulate, apex acuminate, green to purple, exterior surface sericeous, interior glabrous, margin entire. Corolla 1.5–2.2 cm long, 3–4 mm wide at widest point, 2–4 mm wide before limb, 1–2 mm wide at base, tubular, constricted at base, yellow to cream, exterior surface pilose and becoming sericeous with glandular trichomes toward limb, interior glabrate, with a ring of uniseriate trichomes at base, glandular distally; lobes subequal, 1–2 mm long, 1.5–2.5 mm wide, semiorbicular. Filaments connate at base for 3.5–4.5 mm, adnate to base of corolla tube for 1.5–2 mm, white, pilose to apically glabrous; anthers 0.7 mm long, 0.7 mm wide, quadrate, included in corolla tube. Ovary 2–3 mm long, glabrous; style white, glabrous; stigma bilobed, smooth, included in corolla tube. Nectary various, from only 2 dorsal connate enlarged glands to 2 dorsal enlarged and 3 free glands. Berry 6–8.5 mm long, 3–5 mm in diameter, ovoid, pale lavender, glabrous; seeds 1 mm long, 0.3 mm wide, oblong to falcate, twisted, red-brown, striate.

Distribution (Fig. 19). Ecuador; wet forests; 1400–1950 m.

ADDITIONAL SPECIMENS EXAMINED. **Ecuador.** PICHINCHA: Milligalli, near Corazón, *André 3851* (K); 2 km W of Tandápi on new rd Quito–Santo Domingo, *Dodson & Gentry 9592* (MO, SEL); rd Quito–Nanegalito–Pacto, *Smith 1945* (AAU, QCA, QCNE, SEL, US, WIS).

Columnea inconspicua is easily distinguished by its lanceolate to slightly falcate leaves, small inconspicuous pale yellow corolla, narrow calyx lobes, and caducous floral bracts. The last character distinguishes it from *C. manabiana*, its sister species, as revealed by analyses of morphology and cpDNA restriction site variation (Smith & Sytsma 1994a, b, c).

18. Columnea lavandulacea L. P. Kvist & L. E. Skog, Allertonia 6: 387. 1993.—TYPE: ECUADOR. Pichincha: Quito–Santo Domingo rd, 11 Dec 1983, *Kvist & Barfod 49066* (holotype: AAU, photo: AAU!).

Epiphytic or epipetric herbs, usually with climbing or creeping stems 0.3–0.8 m long, 4–6 mm in diameter, terete, red-brown, proximally smooth and glabrous, distally glabrescent to sericeous; internodes 0.6–5.4 cm long; nodes and leaf scars flush with stem. Leaves opposite, anisophyllous; larger leaf laminas 1.7–5.5 cm

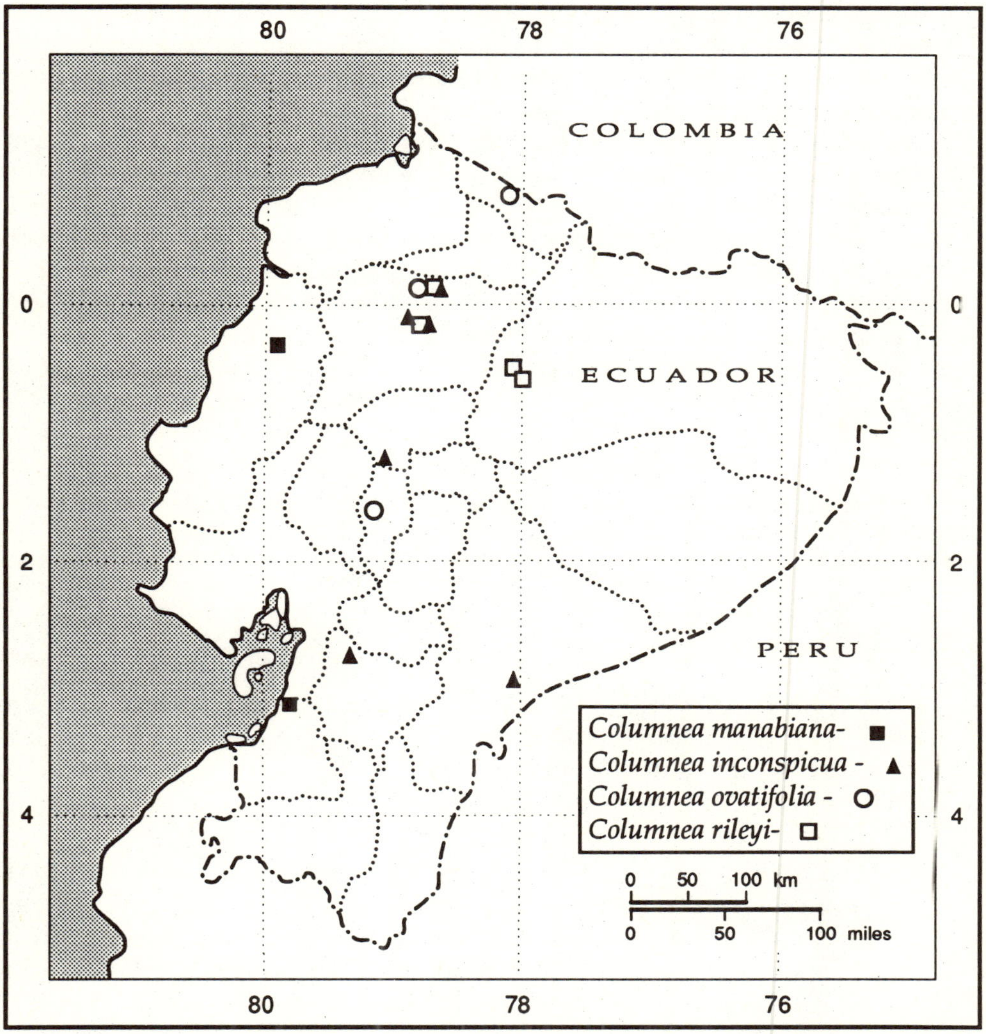

FIG. 19. Distribution of *Columnea manabiana, C. inconspicua, C. ovatifolia,* and *C. rileyi.*

long, 1–2.8 cm wide, ovate to obovate, apex acuminate, base cuneate, oblique, adaxially deep green, glabrous to sparsely pilose, abaxially light green, sparsely pilose to strigose, veins sericeous to sparsely pilose with uniseriate red trichomes, lateral veins 3–6 per side, margin undulate, ciliate with uniseriate red-lavender trichomes; smaller laminas, if present, 0.6–2.6 cm long, 0.25–1.1 cm wide, ovate, otherwise like the larger laminas; petioles 0.5–3 mm long, sericeous. Inflorescences of 1–2, rarely 4, flowers per leaf axil; floral bracts 1–2, 5–11 mm long, 0.3–1.1 mm wide, caducous, linear to lanceolate, green, pilose. Pedicels 4–11 mm long, erect, lavender, sericeous to sparsely pilose, eglandular. Calyx loosely clasping corolla; lobes equal, 10–17 mm long, 1.5–2.0 mm wide, linear, apex acute, dull rose or green, exterior surface appressed-pilose to strigose, interior glabrous, margin entire. Corolla 2.1–2.8 cm long, 4–6 mm wide at widest point, 3–5 mm wide before limb, 1.5–2 mm wide at base, tubular, constricted at base and slightly before limb, lavender, limb yellow, exterior surface pilose with lavender or transparent trichomes, more densely pilose towards limb, interior pilose with glandular hairs distally and dorsally; lobes subequal, 1.5–2.5 mm long, 1.5–2.5 mm wide, semiorbicular, with purple spot on exterior surface. Filaments connate at base for 5–6 mm, adnate to base of corolla tube for 3.5 mm, yellow, basally pilose becoming glabrous distally; anthers 0.8–1.5 mm long, 0.8–1.5 mm wide, quadrate, included in corolla tube. Ovary 2.5 mm long, glabrous; style yellow, glabrous; stigma bilobed, yellow, smooth, included in corolla tube. Nectary of 2 dorsal connate glands. Berry 5–7 mm long, 2.5–3.5 mm in diameter, ovoid, lavender-blue, glabrous; seeds 1.2–1.4 mm long, oblong, twisted, brown, striate. Fig. 20.

Phenology. Flowering continuously.

Distribution (Fig. 5). Northern Ecuador into northern Peru; wet forests; 1540–2430 m.

Additional Specimens Examined. **Ecuador.** Morona-Santiago: 9–10 km SE of San Juan Bosco, *Gentry et al. 30858* (MO, SEL); Cordillera de Cutucú, trail from Logroño to Yaupi, *Madison et al. 3370* (SEL); road from Plan del Milagro to Limón, *Smith 2100* (QCA, WIS).—Pichincha: Mindo, *André K1479* (K, NY); San Florencio, Río Silante, *André 3705* (F, K); between Km 37 and 50 along Río Saloya, *Steyermark 52572* (F, MO, US). **Peru.** Amazonas: 12–18 km E of La Peca in Serrania de Bagua, *Gentry et al. 22920* (MO).

The lavender corollas with purple spots, lavender pubescence, and anisophyllous leaves distinguish *C. lavandulacea* from other species of *Columnea*. It is most similar to the related *C. ovatifolia* (Fig. 2) (Smith & Sytsma 1994a, b, c), which has isophyllous, ovate leaves.

19. Columnea manabiana (Wiehler) J. F. Smith & L. E. Skog, Novon 3: 189. 1993. *Pentadenia manabiana* Wiehler, Phytologia 73: 236. 1992.—Type: cultivated plants from living material (*Dodson & Dodson 6791*), collected in Manabí, Ecuador, Km 67 on rd Chone–Santo Domingo, 500 m, 31 Jul 1977, *Wiehler 87102* (holotype: GES; isotypes: B, F, HBG, K, MO, NY, QCA, SEL, U, US).

Epiphytic herbs with stems to 0.8 m long, 6–8 mm in diameter, squarish when dried, proximally smooth and glabrous, distally pubescent; internodes 1.1–2.6 cm long, slightly swollen; nodes flush with stem; leaf scars raised. Leaves opposite, strongly anisophyllous; larger laminas 8–16.2 cm long, 1.6–4.5 cm wide, lanceolate to slightly falcate, apex acute, base rounded and strongly oblique, adaxially green suffused with pink, abaxially pink-purple, slightly pilose with uniseriate red trichomes on both surfaces, lateral veins 5–6 per side, margin entire to slightly undulate; smaller laminas 1.6–3.7 cm long, 0.4–1.1 cm wide, otherwise like the larger laminas; petioles 2–5 mm long, slightly pilose with uniseriate red trichomes. Inflorescences of 2–4 flowers per axil of the larger leaves; floral bracts 2, 1.3–1.9 cm long, 0.5–0.9 cm wide, conspicuous, ovate, apex acute, green suffused with red, slightly pilose with uniseriate transparent trichomes, margin entire. Pedicels 2.5–4 mm long, villous, eglandular. Calyx loosely clasping corolla; lobes equal to subequal, 1.2–1.6 mm long, 2–4.5 mm wide, lanceolate to elliptic, apex acute, green or green

FIG. 20. *Columnea lavandulacea*. A. Habit. B. Leaf apex. C. Calyx. D. Flower. E. Corolla. F. Corolla interior with retracted stamens. G. Gynoecium with nectaries and calyx (two calyx lobes removed). H. Anthers, adaxial view. I. Anthers, abaxial view. (A, based on *Kvist & Barfod 49066*; B–I, based on *Harling & Andersson 25424*.)

with red tips, slightly pilose with uniseriate transparent trichomes on both surfaces. Corolla 1.8–2.4 cm long, 3–4.5 mm wide at widest point, 2.5–3.5 mm wide before limb, 2–2.3 mm wide at base, tubular, constricted at base, yellow, exterior surface glabrate to villous at apex, interior slightly pubescent at base, distally glandular; lobes inconspicuous, equal to subequal, ca. 1 mm long, 1.5 mm wide, semiorbicular. Filaments connate at base for 5 mm, adnate to base of corolla tube for 4.5 mm, white, slightly pubescent at base; anthers 1.3 mm long, 1.3 mm wide, quadrate, included in corolla tube. Ovary 2 mm long, pilose; style white, slightly pubescent, distally glandular; stigma bilobed, smooth, included in corolla tube. Nectary of 5 free glands, the dorsal glands enlarged. Berry 9 mm long, 5 mm in diameter, ovoid, probably white but dark in color when dried, slightly pilose; seeds not seen. Fig. 21.

Distribution (Fig. 19). Ecuador (Manabí, El Oro); 70–500 m.

ADDITIONAL SPECIMENS EXAMINED. **Ecuador.** EL ORO: hwy from Guayaquil to Machala, 1 km S of Ponce Enriques and 2 km in from rd on low hill, *Dodson et al. 9170* (MO, SEL, US).—MANABÍ: Km 67 on rd from Chone–Santo Domingo, *Dodson & Dodson 6791* (MO).

Columnea manabiana can be distinguished from other species of *Columnea* by the long, narrow, lanceolate to slightly falcate leaves and small yellow corollas shared with *C. inconspicua*, to which it is most closely related (Fig. 2) (Smith & Sytsma 1994a, b, c). It is distinguished from *C. inconspicua* by the presence of large, conspicuous, ovate floral bracts, which partly obscure the inflorescence.

20. Columnea moesta Poeppig, Nov. gen. sp. pl. 3: 1, pl. 201. 1840. *Stygnanthe moesta* (Poeppig) Hanstein, Linnaea 26: 185, 209. 1854. *Pentadenia moesta* (Poeppig) Wiehler, Selbyana 1: 35. 1975.—TYPE: PERU. Ucayali: Pampayacu, *Poeppig 1626* (holotype: W!).

Columnea weberbaueri Mansfeld, Repert. Spec. Nov. Regni Veg. 38: 26. 1935. *Pentadenia weberbaurei* (Mansfeld) Wiehler, Phytologia 27: 315. 1973.—TYPE: PERU. Cuzco, *Weberbauer 7832* (holotype: B, destroyed; lectotype, here designated following L. P. Kvist's annotation: US!; isolectotypes: G! S!).

Small, epiphytic or terrestrial herbs, with frequently branching, suffrutescent to scandent sublignose stems, 60–80 cm tall, 2.5–6.0 mm in diameter, square, red-brown, proximally smooth and glabrous, distally hirsute with uniseriate red or transparent trichomes; internodes 0.2–8.6 cm long; nodes flush with stem; leaf scars raised. Leaves opposite, slightly anisophyllous, frequently clustered at apex; laminas 2.4–10.2 cm long, 1.5–4.0 cm wide, elliptic, apex acute to acuminate, base rounded to cuneate, adaxially green, strigose to appressed-pilose to hirsute with uniseriate transparent or red or single-celled transparent trichomes, abaxially green-pink, appressed-pilose with uniseriate transparent or red trichomes, more densely pilose on veins, lateral veins 5–6 per side, margin minutely crenulate, ciliate with red uniseriate trichomes; petioles nearly absent or to 27 mm long, appressed-pilose with uniseriate red or transparent trichomes. Inflorescences of a solitary flower per axil; floral bract ca. 1, 3.5 mm long, 0.5 mm wide, linear, red, pilose, caducous. Pedicels 1.0–1.5 (–2.8) cm long, erect in axil, hirsute to villous, glands conspicuous, 0.8 mm long, oblong, dark purple. Calyx clasping corolla; lobes equal to subequal, 1–18 mm

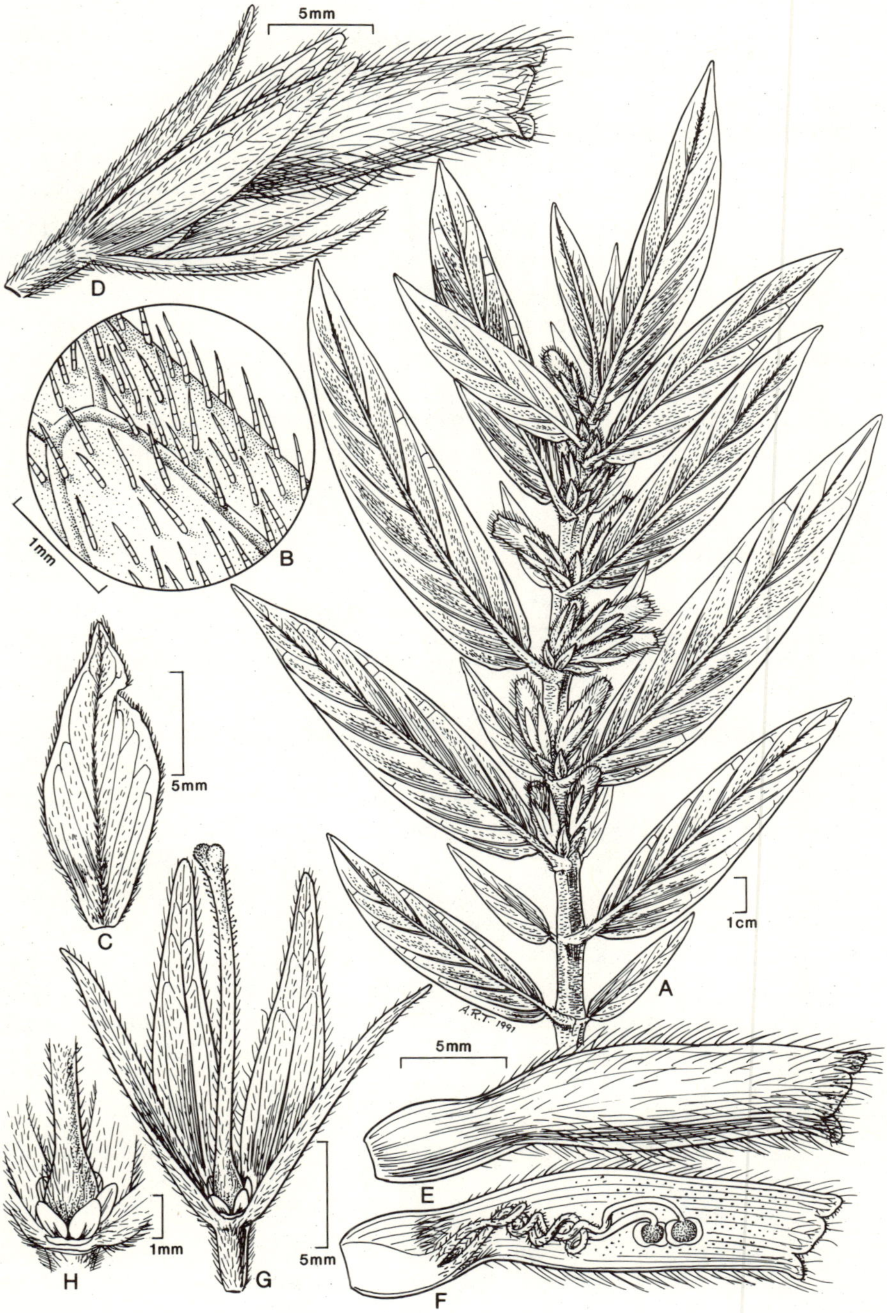

FIG. 21. *Columnea manabiana*. A. Habit. B. Abaxial leaf pubescence. C. Inflorescence bract. D. Flower. E. Corolla. F. Corolla interior with retracted stamens. G. Gynoecium with calyx and nectaries (one calyx lobe removed). H. Base of gynoecium with nectaries (one calyx lobe removed). (A–B, based on *Dodson et al. 9170*; C–H, based on *Dodson et al. 6791*.)

long, 1–3.5 mm wide, lanceolate to linear, apex acute, green or with red tips, pilose with uniseriate transparent trichomes on both surfaces, margin denticulate to entire. Corolla 3.3–4.5 cm long, 7–15 mm wide at widest point, 4–7 mm wide before limb, 2–6 mm wide at base, tubular, ventricose, constricted at mouth, red-violet, limb green, exterior surface long-villous, more densely villous toward limb, interior hirsute at base with capitate-glandular trichomes dorsally and distally; lobes equal, 1.5–2.5 mm long, 1.5–4.5 mm wide, suborbicular, with purple spots on interior surface. Filaments connate at base for 0.7 mm, pale pink-purple, slightly pilose to hirsute; anthers 1.5 mm long, 0.8 mm wide, rectangular, included in corolla tube. Ovary 3 mm long, pilose; style pale pink, proximally slightly pilose to hirsute, distally glandular; stigma bilobed, yellow, smooth, included in corolla tube. Nectary of 5 free glands. Berry 1.2 cm in diameter, globose, glabrous to slightly pubescent, violet; seeds ca. 1 mm long, fusiform, twisted, light brown, striate.

Phenology. Flowering from October to February, fruiting in April.

Distribution (Fig. 16). Southern Peru and northern Bolivia; wet forests and cloud forests; 600–2450 m.

ADDITIONAL SPECIMENS EXAMINED. **Bolivia.** COCHABAMBA: Prov. Chaparé, Valle de Corani, unknown collector *459* (US); rd from Cochabamba to Villa Tunari, *Besse et al. 563* (SEL), *Luer et al. 3535*, *4936*, *4890* (SEL); Km 94 on rd from Cochabamba to Villa Tunari, near Hotel Caballeros, *Croat 51361* (MO), *Smith 1776* (LPB, WIS); Incachaca, *Steinbach 8929* (GH, MO, NY, S). **Peru.** CUZCO: near Río Yanamayo, below "Pillahuata," *Pennell 14038* (PH); Quispicanchis, Hda. Río Marcapata, *Vargas C. 3073* (US).

Columnea moesta is similar to *C. trollii* but is distinguished by its smaller leaves with shiny, silky vesture and a corolla that is narrower, red-violet with green limb and purple spotted lobes, as opposed to a red-orange corolla with yellow limb. Collections from Peru, originally described as *C. weberbaueri*, are clearly specimens of *C. moesta* but generally of smaller stature and with smaller, narrower corollas and leaves.

Analysis of cpDNA evidence (Smith & Sytsma 1994b) suggests that *C. moesta* is related to *C. ultraviolacea* of sect. *Stygnanthe*, but cladistic analysis of morphological data (Smith & Sytsma 1994a) allies it with *C. trollii*. Such conflicts in phylogenetic analyses frequently indicate a hybrid accession; however, analysis of biparentally inherited nuclear ribosomal DNA, which complements the uniparentally inherited cpDNA, indicates that the accession is not a hybrid and is allied with *C. ultraviolacea* (Smith 1991). In addition, no morphological intermediates between the two species were found in herbarium material examined. Although both species are from Bolivia, they are not from the same locality, and the possibility of hybridization is low. Because of the larger number of characters (12 vs. 3) that place *C. moesta* with *C. ultraviolacea* rather than with *C. trollii*, the analysis of the combined data favors the placement as seen with cpDNA (Fig. 2) (Smith & Sytsma 1994c). Therefore, despite its morphological similarity to some species of sect. *Pentadenia*, *C. moesta* is placed in sect. *Stygnanthe*.

21. Columnea orientandina (Wiehler) L. P. Kvist & L. E. Skog, Allertonia 6: 392. 1993. *Pentadenia orientandina* Wiehler, Selbyana 2: 123. 1977.—TYPE: cultivated material, grown from cuttings (*Madison & Coleman 2532*) collected in the Cordillera de Cutucú, Morona-Santiago, Ecuador, *Wiehler 77123* (holotype: SEL!; isotype: US!).

Epiphytic, rarely terrestrial herbs with ascending or spreading stems 0.3–0.7 m long, 4–6 mm in diameter, terete, green or tawny, proximally smooth, slightly hirsute to glabrous, distally hirsute to sericeous with red uniseriate trichomes; internodes 0.4–3.3 cm long, slightly swollen; nodes and leaf scars flush with stem. Leaves opposite, strongly anisophyllous; larger laminas 2.2–8.2 cm long, 0.9–3.2 cm wide, elliptic to oblanceolate, apex obtuse, base rounded, strongly oblique, adaxially green, pilose to hirsute, abaxially with red apex, sericeous to lanate, appressed-sericeous on veins, lateral veins 4–8, margin entire; smaller laminas, if present, 0.6–2.7 cm long, 0.3–1.5 cm wide, otherwise like the larger laminas; petioles 1–5 mm long, green, sericeous. Inflorescences of 2–8 flowers in axils of larger leaves; floral bract 1, 3–5 mm long, 0.5–1.0 mm wide, linear, apex acute, green, sericeous. Pedicels 4–10 mm long, green or maroon, sericeous, eglandular. Calyx loosely clasping corolla; lobes equal to subequal, 6–9 mm long, 0.4–1.0 mm wide in flower, enlarging to 8–13 mm long, 1–1.8 mm wide in fruit, lanceolate to subulate, acuminate, green in flower, tips becoming red in fruit, exterior surface sericeous, interior glabrous, margin subentire to minutely denticulate in fruit. Corolla 1.7–2.5 cm long, 3–6 mm wide at widest point, 2–4 mm wide before limb, 1–2 mm wide at base, tubular, constricted at base, lemon-yellow, exterior surface puberulent becoming long-glandular-pilose toward limb, interior glabrate with a ring of trichomes at base, glandular distally; lobes subequal, inconspicuous, 1.0–1.5 mm long, 1.0–2.0 mm wide, semiorbicular. Filaments connate at base for 3.5 mm, adnate to base of corolla tube for 2 mm, white, glabrous; anthers 0.7–1.0 mm long, 0.7–1.0 mm wide, quadrate, slightly exserted. Ovary 2 mm long, green, sericeous; style white, proximally glabrous, distally pilose; stigma stomatomorphic, papillose, slightly exserted. Nectary varying from 5 glands with the 2 dorsal connate to 4 free glands. Berry 1.7 cm long, 1 cm in diameter, ovoid, pink, pilose; seeds 1.3 mm long, oblong, twisted, striate.

Distribution (Fig. 8). Ecuador (Morona-Santiago) and Peru (Pasco); 1000–1500 m.

ADDITIONAL SPECIMENS EXAMINED. **Ecuador.** MORONA-SANTIAGO: Cordillera de Cutucú, along trail from Logroño to Yaupi, *Madison et al. 3420* (SEL); Cordillera de Cutucú, 25 km SE of Logroño, *Madison & Coleman 2537* (SEL).—NAPO: Puyo–Tena rd, near Santa Clara, *Kvist et al. 60352* (US). **Peru.** PASCO: Pichis trail, San Nicolas, *Killip & Smith 26059* (NY).

Columnea orientandina is readily distinguished from other species of *Columnea* by its relatively compact habit, laminas with a blunt red apex, and yellow corollas. It is similar to *C. inconspicua*, which has green, acuminate laminas, and *C. byrsina*, which has bright red corollas. The exserted anthers and stigma of *C. orientandina* are also distinguishing characters, but these are not seen in all specimens.

Cladistic analyses of morphology and cpDNA variation indicate that *C. orientandina* is the sister species to *C. byrsina* and is also related to *C. colombiana* (Fig. 2) (Smith & Sytsma 1994a, b, c).

22. Columnea ovatifolia L. P. Kvist & L. E. Skog, Allertonia 6: 393. 1993.—TYPE: ECUADOR. Carchi: rd Tulcán–Maldonado, 10 km SE of Maldonado, Campamente Machines, 28 Nov 1974, *Harling & Andersson 12316* (holotype: GB!; isotype: SEL).

Epiphytic herbs with pendent and creeping stems to 50 cm long, 2–5 mm in diameter, terete, red-brown, proximally smooth, glabrous, distally glabrescent to sericeous with violet uniseriate trichomes; internodes 1.1–6.5 cm long; nodes and leaf scars flush with stem. Leaves opposite, isophyllous to slightly anisophyllous; laminas 1.5–2.5 cm long, 1.0–1.7 cm wide, ovate to orbicular, apex acute to obtuse, base rounded, slightly oblique, adaxially dull green, often with violet spots, glabrous to pilose, abaxially lighter green, sparsely pilose to strigose, veins appressed sericeous, lateral veins 3–4, margin entire, ciliate with uniseriate violet trichomes; petioles 1–4 mm long, sericeous to pilose. Inflorescences of a solitary flower per axil; floral bracts 1–2, 1.5 mm long, caducous, sericeous. Pedicels 3–6 mm long, erect, sericeous to pilose with lavender uniseriate trichomes, glands near calyx 0.6 mm long, round to oval. Calyx loosely clasping corolla; lobes equal, 11–15 mm long, 2–3 mm wide, linear, acuminate, green flushed with red, exterior surface pilose to strigose, interior glabrous, margin entire. Corolla 4.0–4.8 cm long, 5–8 mm wide at widest point, 4–5 mm wide before limb, 2.0–2.5 mm wide at base, tubular, ventricose, with a long constriction at base, slightly constricted at limb, pink to red-violet, limb pale yellow, exterior surface slightly pilose to sericeous, interior slightly pilose to glabrous, glandular dorsally and distally; lobes subequal, 1.5–2.5 mm long, 1.5–2.5 mm wide, semiorbicular, with dark purple spots on interior surface. Filaments connate at base for 4–7 mm, adnate to base of corolla tube for 3–5 mm, proximally pilose, becoming glabrous distally; anthers 1.0–2.5 mm long, 1.0–2.5 mm wide, quadrate, included in corolla tube. Ovary 3–5 mm long, densely pilose; style yellow, sparsely pilose to glabrous; stigma bilobed, yellow, papillose, included in corolla tube. Nectary of 2 dorsal connate glands and 3 ventral connate glands. Fruit and seeds not seen. Fig. 22.

Phenology. Flowering from January to May.

Distribution (Fig. 19). Northern Ecuador; cloud forests; 2380–2800 m.

ADDITIONAL SPECIMENS EXAMINED. **Ecuador.** BOLÍVAR: Atio de Telimbela, *Acosta S. 7160* (F).—PICHINCHA: 9 km S of Tandayapa towards Mindo, *Luteyn et al. 8830* (NY); rd Quito–Tandayapa–Mindo, *Smith 1921* (QCA, QCNE, WIS).

Columnea ovatifolia is most similar to the related *C. crassicaulis* (Fig. 2) (Smith & Sytsma 1994a, b, c), but can be distinguished from this species by its smaller leaves, pink-purple corolla, and thinner stem.

23. Columnea rileyi (Wiehler) J. F. Smith, comb. nov. *Pentadenia rileyi* Wiehler, Phytologia 73: 236. 1992.—TYPE: ECUADOR. Napo: 37 km from Baeza on rd to Lago Agrio, 24 Apr 1986, *Wiehler & GRF Expedition 86243* (holotype: GES; isotypes: F, K, MO, NY, QCA, SEL, U, US).

Columnea leucerinea L. P. Kvist & L. E. Skog, Allertonia 6: 389. 1993.—TYPE: ECUADOR. Napo: Lago Agrio–Baeza rd, Km 145, Río Aya Cachi, 8 Jan 1987, *Kvist et al. 60377* (holotype: AAU; isotypes: COL, MO, NY, QCA, QCNE, US!)

Epiphytic or terrestrial, herbaceous shrubs, branching at base, stems 40–100 cm long, 3–6 mm in diameter, terete, green, sericeous to lanate; internodes 1.5–3.8 cm long; nodes and leaf scars flush with stem. Leaves opposite, isophyllous to slightly anisophyllous; laminas 3.7–8.0 cm long, 1.0–2.5 cm wide, lanceolate to ovate or

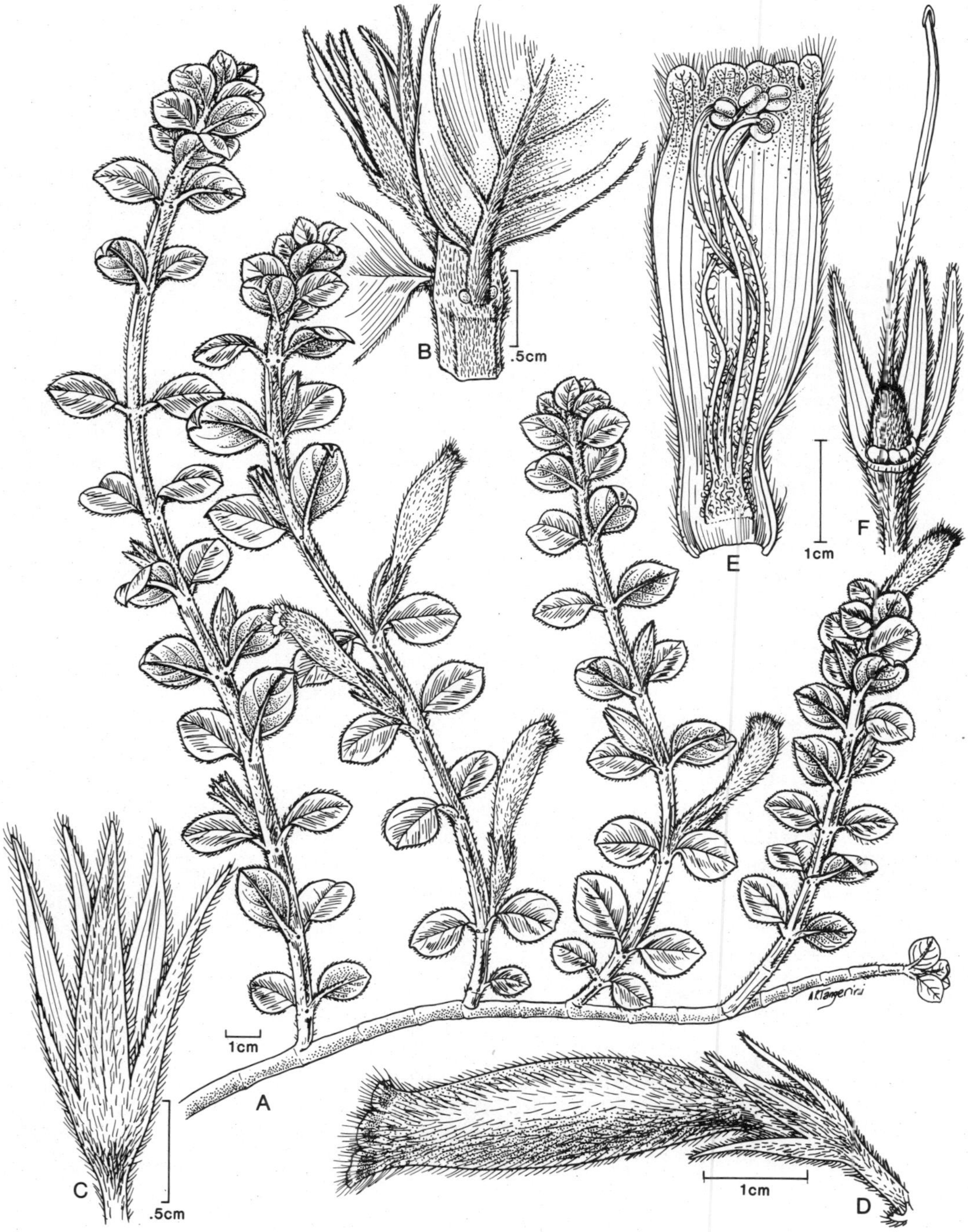

FIG. 22. *Columnea ovatifolia*. A. Habit. B. Node showing adventitious root initials. C. Calyx. D. Flower. E. Corolla interior with stamens. F. Gynoecium with nectaries (two calyx lobes removed). (Based on *Harling & Andersson 12316* and *Luteyn et al. 8830*.)

elliptic, apex acuminate, base cuneate, oblique, adaxially dull green, pubescent to tomentose, abaxially green, sericeous to lanate, veins sericeous to lanate, lateral veins 4–5, margin serrulate, ciliate with uniseriate red or transparent trichomes; petioles 5–15 mm long. Inflorescences of 1–4 flowers per axil; floral bracts (1–) 3–5, 3–8 mm long, caducous, linear, green, sericeous to lanate. Pedicels 4–15 mm long, erect, lanate, eglandular. Calyx loosely clasping corolla; lobes equal, 7–12 mm long, 1–4 mm wide, spatulate to oblanceolate, apex acute, maroon or green with purple tips, exterior surface sericeous to lanate, interior glandular-pubescent, margin entire. Corolla 1.4–2.1 cm long, 3–5 mm wide at the widest point, 1.5–2.5 mm wide before limb, 2–3 mm wide at base, tubular, constricted at base, orange-red, exterior surface villous, with red uniseriate trichomes, interior glabrous; lobes subequal, 1.5 mm long, 1.2 mm wide, semiorbicular, with dark orange spot on interior surface. Filaments connate at base for 4–6 mm, adnate to base of corolla tube for 1.5–2.0 mm, glabrous to sparsely pilose; anthers 1.2 mm long, 1 mm wide, quadrate, included in corolla tube. Ovary glabrous; style white, glabrous; stigma stomatomorphic, smooth, included in corolla tube. Nectary of 5 free glands. Berry 8 mm in diameter, globose, white, glabrous; seeds 1.1 mm long, oblong, twisted, light brown, striate. Fig. 23.

Distribution (Fig. 19). Northern and western Ecuador; wet forests; 1500–2000 cm.

ADDITIONAL SPECIMENS EXAMINED. **Ecuador**. PICHINCHA: rd Quito–Nanegalito–Pacto, *Smith 1944* (QCA, QCNE, WIS).

The dense white pubescence covering the entire plant body of *C. rileyi* is sufficient to distinguish it from any other species of *Columnea*. *Columnea rileyi* could be confused only with *C. lophophora*, another species with dense vesture; however, the small orange-yellow corollas and small floral bracts easily distinguish *C. rileyi* from *C. lophophora*.

Cladistic analysis of cpDNA variation (Fig. 2) (Smith & Sytsma 1994b) shows *C. rileyi* is closely related to *C. lavandulacea*, *C. ovatifolia*, and *C. crassicaulis*, although cladistic analysis of morphology places it with *C. colombiana* (Smith & Sytsma 1994a). *Columnea rileyi* is allied with *C. colombiana* in the morphological analysis on the basis of only two convergent characters, and thus this placement is not as well supported as its position in the cpDNA analysis.

24. Columnea spathulata Mansfeld, Notizbl. Gard. Berlin-Dahlem 14(121): 37. 1938. *Pentadenia spathulata* (Mansfeld) Wiehler, Phytologia 27: 315. 1973.—TYPE: ECUADOR. Pichincha: Santo Domingo de los Colorados, *Schultze-Rhonhof 1876* (holotype: B, destroyed).—ECUADOR. Pichincha: Santo Domingo de los Colorados, Centinella, Montañas de Ila, 12 km from Patricia Pilar, 575 m, 10 Jul 1979, *Løjtnant & Molau 15811* (neotype, designated by Kvist & Skog, 1993: AAU; isoneotype: GB!).

Alloplectus microsepalus C. Morton, Fieldiana, Bot. 28: 523. 1953. *Pentadenia microsepala* (C. Morton) Wiehler, Phytologia 27: 375. 1973. *Columnea microsepala* (C. Morton) L. P. Kvist & L. E. Skog, Allertonia 6: 391. 1993.—TYPE: VENEZUELA. Monagas: Cerro de la Cueva de Doña Anita, S of and bordering valley of Caripe, 1100–1200 m, 7 Apr 1945, *Steyermark 61905* (holotype: F!; isotype: US!).

FIG. 23. *Columnea rileyi*. A. Habit. B. Calyx lobes. C. Flower. D. Corolla interior with stamens. E. Gynoecium with nectaries and calyx. F. Stigma. G. Berry (one calyx lobe removed). H. Seeds. (A–B, D–H, based on *Holm-Nielsen et al. 26229*; C, based on *Kvist et al. 60377*.)

Pentadenia zapotalana Wiehler, Selbyana 2: 85, pl. 26B. 1977. *Columnea zapotalana* (Wiehler) L. E. Skog, Taxon 33: 126. 1984.—TYPE: ECUADOR. Los Ríos: 20 km S of Quevedo, *Wiehler et al. 71312* (holotype: SEL!; isotype: US!).

Epiphytic or epipetric herbs, with vining or upright stems to 1 m, 4–15 mm in diameter, occasionally with zigzag appearance, terete, proximally smooth and glabrous, distally glabrate to villous, frequently with conspicuous adventitious roots; internodes 0.4–8.6 cm long; nodes and leaf scars flush with stem. Leaves opposite, strongly anisophyllous, rarely slightly anisophyllous; larger laminas 3.5–14 cm long, 1.6–6.0 cm wide, oblanceolate to elliptic, rarely ovate, apex acute to obtuse, base rounded, strongly oblique, adaxially green to purple, pilose to villous with uniseriate transparent or red trichomes, abaxially green, green with red-purple apex, green with red-purple mottling, or entirely red-purple, vesture as on adaxial surface, lateral veins 4–6 per side, margin crenate, ciliate with red uniseriate trichomes; smaller laminas 6–18 mm long, 3–12 mm wide, rarely to 6.0 cm long, 2.3 cm wide, ovate to elliptic, otherwise like the larger laminas; petioles absent to 16 mm long. Inflorescences of 1–12 flowers per axil of the larger leaves, rarely in both axils of a leaf pair; floral bract frequently absent or 1, 2–8 mm long, 2–7 mm wide, lanceolate to ovate, apex acute, green, green with red tips, pink, or red, pilose to villous, margin entire. Pedicels absent to 8 mm long, villous to sericeous, eglandular. Calyx either clasping corolla or free from it; lobes equal to subequal, 5–15 mm long, 1–6 mm wide, spatulate to lanceolate, apex acute to obtuse, green, green with purple tips, or entirely purple, villous to hirsute, sericeous at base with uniseriate transparent trichomes, margin entire. Corolla 1.0–2.8 cm long, 1.5–5.0 mm wide at widest point, 1.5–5.0 mm wide before limb, 1.0–2.5 mm wide at base, tubular, constricted at base, yellow, orange, or red, exterior surface pubescent to pilose with uniseriate red or transparent trichomes, villous towards limb, interior pilose to villous with uniseriate red or transparent trichomes, distally glandular; lobes equal, inconspicuous, 1–2 mm long, 1.2–2.0 mm wide, semiorbicular. Filaments connate at base for 2.0–4.5 mm, adnate to base of corolla tube 1.0–3.5 mm, white, proximally slightly pilose, distally glabrous; anthers 0.5–1.0 mm long, 0.5–1.0 mm wide, quadrate, included in corolla tube. Ovary 0.8–2.0 mm long, pubescent to sericeous with uniseriate red or transparent trichomes; style white, slightly pubescent to pilose, glandular at apex; stigma bilobed, smooth, included in corolla tube. Nectary of 5 free glands, sometimes with the 2 dorsal glands connate. Berry 7–10 mm in diameter, globose, white, glabrate; seeds 0.8–1.2 mm, oblong, twisted, light brown to red-brown, striate.

Phenology. Flowering and fruiting continuously throughout most of its range, restricted to July to April in Venezuela.

Distribution (Fig. 24). Venezuela to Bolivia; cloud forest, wet forest, also in disturbed areas, such as *Citrus* plantations; 75–2800 m.

REPRESENTATIVE SPECIMENS. **Venezuela.** AMAZONAS: Sierra Parima, 45 km NW of the head of the Río Orinoco, *Steyermark 105926* (F).—ANZOÁTEGUI: Dtto. Bolívar, Fila El Guácharo, above Los Chorros, and El Cielo, Serranía de Turimiquire, *Davidse & González 19450* (MO).—DISTRITO FEDERAL: rd between Portachuelo and Peñita between Colonia Tovar–Junquito rd and El Limón, 6 mi below junction of Junquito–Colonia Tovar rd, *Steyermark 90952* (US).—FALCÓN: Sierra de San Luis,

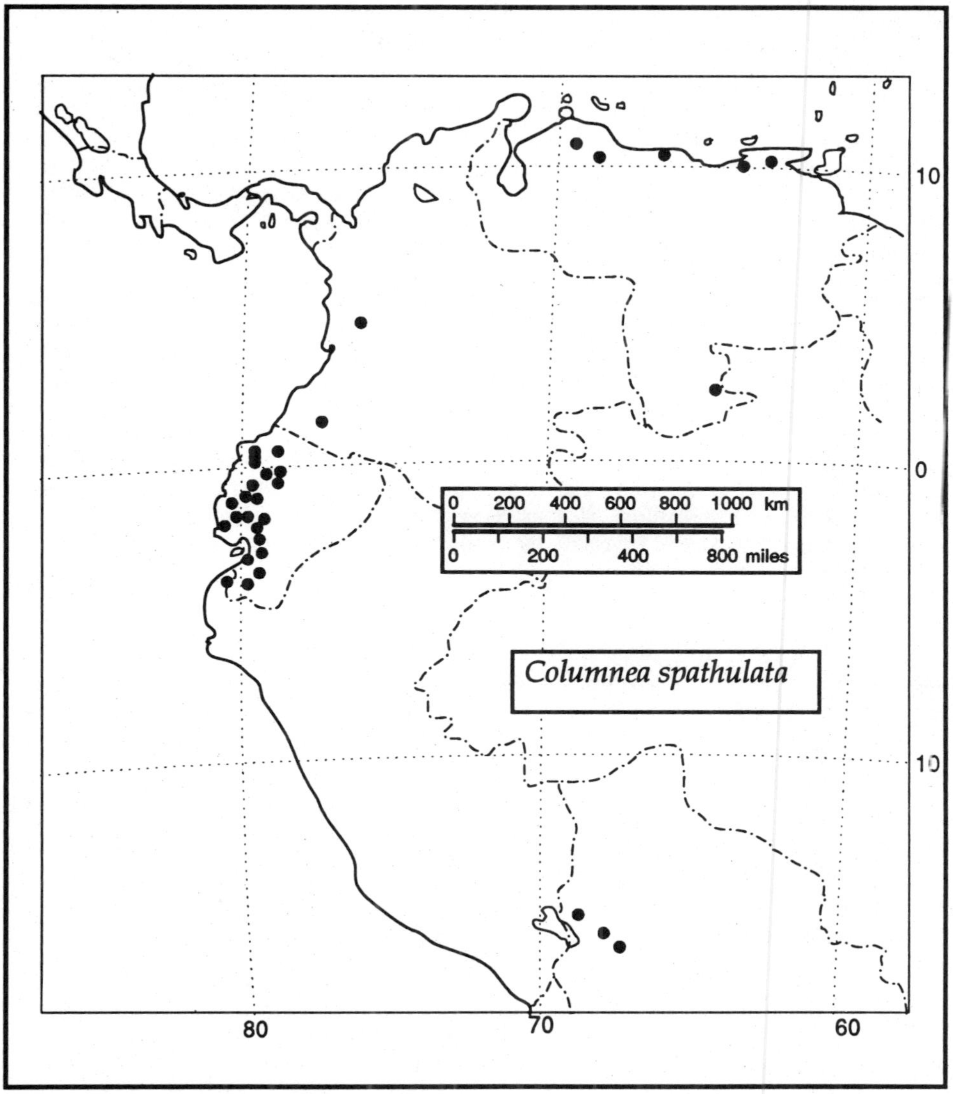

FIG. 24. Distribution of *Columnea spathulata*.

Paraguariba, 3 km ENE Cerro Galicia, 2–3 km NW of Curimagua, *Plowman et al. 13411* (F); rd from Curimagua to San Luis, Guaritaro, *Smith 1221* (WIS); Sierra de San Luis, near Hotel Parador, *Steyermark 98976* (WAG); up from La Chapa, unknown collector *193* (US).—MONAGAS: 4 km up from junction with Caripe–Teresán rd, S side of Cerro San Bonifacio in Bella Vista, *Bunting 2637* (BH).—YARACUY: Salome area, cultivated plant at SEL, *Wiehler 77135* (SEL). **Colombia.** CHOCÓ: rd San Jose del Palmar–Névita, Campamente Curundé, Río Ingará, *Forero et al. 2322* (COL, MO, US).—NARIÑO: Benavida Finca, Km 63 on Tumaco–Tuquerres rd, *Gentry et al. 34920*. **Ecuador.** AZUAY: 12 km from Guayas border, *Boeke & Loyola 2176* (NY, SEL, US); rd Pasaje–Santa Isabel–Girón, *Harling & Andersson 14460* (GB, SEL).—BOLÍVAR: Atio de Telimbela, *Acosta S. 6885* (F).—COTOPAXI: Tenefuerte, Río Pilalo, Km 52–53, Quevedo, Latacunga, *Dodson & Gentry 12791* (MO, SEL); trail El Corazón–Ramón Campaña, 1–3 km N of El Corazón, *Harling & Andersson 19245* (GB); Quevedo–

Latacunga rd, *Holm-Nielsen et al. 2972* (S).—EL ORO: Malva, 8 km NW of Zaruma, *Dodson et al. 8434* (SEL); Km 19 from Piñas to Sta. Rosa, *Dodson et al. 8891* (MO, SEL); 11 km W of Piñas, *Dodson et al. 9138* (MO, SEL, US); between Piñas and El Placer, *Harling & Andersson 14367* (GB).—ESMERALDAS: E of Río Blanco, S of Quininde, *Fagerlind & Wibom 2595* (S); 20 km SE of Viche, ca. 56 km SE of Esmeraldas on rd to Santo Domingo, *Hansen et al. 7953* (SEL).—GUAYAS: 5 km N of Manglaralto on beach, *Dodson & Thien 1674* (US).—IMBABURA: valley of Río Mira, between Ibarra and Lita, 2.5 km E of Lita, *Croat 38969* (MO).—LOJA: rd from Alamor–El Limo, just before El Limo, *Smith 1960* (QCA, QCNE, WIS).—LOS RÍOS: Km 21 Quevedo–Santo Domingo, near Buena Fe, *Dodson 5974* (SEL); roadside at Quevedo, *Haught 2965* (BH, US); 58 km ENE of Quevedo on Pan Am hwy to Santo Domingo, *Iltis et al. E-59* (WIS); Nuevo Zapotal, 2 km S of Quevedo, *Wiehler et al. 71312* (SEL, US).—MANABÍ: slopes of Montecristi, *Besse et al. 1272* (SEL).—PICHINCHA: Mindo, *André K1466* (K); 9.6 km E of Santo Domingo on rd to Aloag, near Hotel Tinalandia, *Croat 55658* (MO, US); Paraíso, 3 km W of Alluriquim on rd Santo Domingo–Quito, *Dodson & Gentry 9577* (MO, SEL); Finlandia, 16 km E of Santo Domingo, *Gentry et al. 12164* (MO, US); 11 km W of Santo Domingo on rd to Chone, 2 km S of Finca La Carlita, *Iltis & Iltis E-276* (SEL, WIS); Hotel Zaracay, Santo Domingo, *Játiva & Epling 1183* (NY, S, US); old rd Quito–Santo Domingo, *Smith 1853* (QCA, QCNE, WIS), *Smith 1900* (QCA, QCNE, WIS); rd Santo Domingo–Aloag, confluence between ríos Toáchi and Pilatón, *Sparre 17825* (S). **Peru.** TUMBES: Prov. Zarumilla: Bosque Nacional de Tumbes, near Campe Verde, *Simpson & Schunke V. 430* (F), *461* (F, US). **Bolivia.** LA PAZ: Prov. Munecas, along Río Cristóbal, near Consata, *Besse et al. 597* (SEL); Prov. Murillo, 45.5 km below dam at Lago Zongo, *Solomon 12896* (MO, US); Prov. Nor Yungas, 10.2 km SE of Yolosa on rd to Chuspipata, *Solomon & Uehling 12232* (US).—Locality unknown: *Bogner 944* (M).

Columnea spathulata is distinctive in sect. *Stygnanthe.* It has strongly anisophyllous leaves and a fasciculate inflorescence of many flowers. Several other species of sect. *Stygnanthe* have strongly anisophyllous leaves but none the dense inflorescence of *C. spathulata*. In addition, the crenate laminar margin distinguishes this species from others of sections *Pentadenia* and *Stygnanthe*.

Columnea spathulata is highly variable. Leaf coloration can range from all-purple to all-green. Corolla color is generally yellow, although some collections from Ecuador, and those of Peru, have red or orange corollas. The number of flowers per axil also ranges widely from 1 to 12; they are organized into a very compact or open inflorescence. In general, specimens from Venezuela, Peru, and Bolivia have all-green vegetative parts and tend to have a robust appearance, whereas specimens from Colombia and northern Ecuador generally have all-purple leaves and dense inflorescences, and the collections from southern Ecuador have green leaves with red apices or red mottling below and an open inflorescence of fewer flowers.

The type specimens of the three names applied to *C. spathulata* are distinctive and have long been thought to merit species status. Yet, examination of collections from throughout the range reveals numerous specimens that are intermediate between each pair of variants, and others that are similar but with subtle differences. Some of these intermediates may represent hybrids between the different morphologies; however, the geographic distribution of the different intermediates does not correspond well to overlapping zones of the putative parental species. Therefore, it is more likely that the intermediates represent localized morphological variants. Because the variation appears to be continuous, all variants are here considered to represent a single species. In addition, analysis of cpDNA restriction site variation indicates that only a single mutation occurs among one accession representing "*C. zapotalana*" and two accessions, each representing *C. spathulata* s.str. and "*C. microsepala*" (Smith & Sytsma 1994b).

Cladistic analysis of cpDNA data and combined analysis of morphology and cpDNA variation indicate that *C. spathulata* is most closely related to *C. angustata*, another widespread species (Fig. 2) (Smith & Sytsma 1994b, c). Based on morphology alone, it is basal to a polytomy containing most of the species of sect. *Stygnanthe* and the species representing sections *Columnea*, *Collandra*, and *Ortholoma* (Smith & Sytsma 1994b). Because the larger number of cpDNA characters ally *C. spathulata* with *C. angustata*, and the low number of morphological characters place it as basal to the polytomy, the combined analysis favors the position of the cpDNA analysis (Smith & Sytsma 1994a, b, c).

Columnea spathulata, distributed from Venezuela to Bolivia, is the most widespread species in sect. *Stygnanthe*. Most species of *Columnea* are narrow endemics (Kvist & Skog 1993), and only a few have broad ranges, even though they are all berry-fruited and bird-dispersed. *Columnea spathulata* obviously possesses characteristics that allow for its survival in a wide range of habitats, as indicated by its geographic and altitudinal distribution. In addition, *C. spathulata* is frequently collected in cacao and citrus plantings, which suggests tolerance of disturbance.

25. Columnea suffruticosa J. F. Smith & L. E. Skog, Novon 3: 190. 1993.—TYPE: COLOMBIA. Chocó: Mpio. San José del Palmar, Cerro del Torrá, 7 Jan 1984, *Silverstone-Sopkin et al. 1594* (holotype: CUVC!; isotypes: MO! US!).

Small, woody to sublignose, epiphytic or terrestrial shrubs with stems to 2 m, 2–3 mm in diameter, terete, proximally dark purple, smooth, glabrous, distally pubescent or hirsute with uniseriate trichomes; internodes 1.0–3.5 cm long; nodes and leaf scars flush with stem. Leaves opposite, anisophyllous; larger laminas 1.4–5.0 cm long, 0.9–2.5 cm wide, elliptic to ovate, apex acute, base rounded to oblique, adaxially dull green, glabrate, abaxially purple to mottled to all-purple, glabrate, lateral veins 3–6 per side, margin crenulate; smaller laminas 0.7–3.9 cm long, 0.4–1.8 cm wide, elliptic to ovate, otherwise like the larger laminas; petioles 2–10 mm long, pilose to hirsute. Inflorescences of 1–2 flowers per axil; floral bract 1, up to 3 mm long, caducous, linear. Pedicels 5–11 mm long, erect, hirsute with purple or transparent uniseriate trichomes, eglandular. Calyx loosely clasping corolla; lobes equal to subequal, 6–11 mm long, 1.5–5.0 mm wide, spatulate, acute, green or purple, exterior surface hirsute to slightly so, interior glabrous, margin serrate. Corolla 1.5–3.1 cm long, 4–6 mm wide at widest point, 3–5 mm wide before limb, 2 mm wide at base, tubular, constricted at base, proximally and ventrally with two small invaginations of corolla tube, yellow, exterior surface slightly sericeous with purple or transparent uniseriate trichomes, interior glabrous, glandular distally and dorsally; lobes subequal, 1–2 mm long, 1.5–2.0 mm wide, semiorbicular, with orange spots on interior surface. Filaments connate at base for 2.5 mm, adnate to base of corolla tube for 1.5 mm, white, glabrous; anthers 1.5 mm long, 1.5 mm wide, quadrate, included in corolla tube. Ovary 2 mm long, sericeous at apex; style white, glabrous with glandular trichomes distally; stigma stomatomorphic, white, smooth, included in corolla tube. Nectary of 5 free glands. Mature fruit unknown; immature berry ovoid, white, sparsely pilose; seeds not seen. Fig. 25.

Distribution (Fig. 18). Colombia (Chocó, Valle del Cauca); 1870–2500 m

FIG. 25. *Columnea suffruticosa*. A. Habit. B. Adaxial leaf pubescence. C. Flower. D. Corolla interior with stamens and gynoecium. E. Pouchlike protuberances on interior of corolla. F. Gynoecium with nectaries and calyx. G. Berry with nectaries. (A, based on *Silverstone-Sopkin et al. 1594*; B, based on *Silverstone-Sopkin et al. 4289*; C–G, based on *Silverstone-Sopkin et al. 4530*.)

ADDITIONAL SPECIMENS EXAMINED. **Colombia.** CHOCÓ: Mpio. San José del Palmar, Cerro del Torrá, up from heliport, *Ramos et al. 1357* (CUVC), *Silverstone-Sopkin 1225* (CUVC), *1496* (CUVC, MO), *1563* (MO), *1548* (MO), *1812* (CUVC), *4289* (CUVC, US), *4442* (CUVC), *4530* (CUVC, US).—VALLE DEL CAUCA: Mpio. El Cairo, ca. 21–25 km beyond El Cairo toward Chocó border, *Luteyn et al. 12288*, *12644* (CUVC).

The shrubby woody habit of *C. suffruticosa* readily distinguishes it from other species of *Columnea*. The small, ovate, crenate leaves are also distinctive. *Columnea suffruticosa* is most likely related to *C. colombiana*, as shown by a cladistic analysis of morphology (Smith & Sytsma 1994b); however, the lobes with darker colored spots, a character not used in the analysis, may place it in the clade with *C. ovatifolia, C. lavandulacea*, and *C. crassicaulis*.

Two of the four collections studied are epiphytic; the others are terrestrial. The subtle morphological differences between the two types are most likely due to environmental effects of the two habitats. The epiphytic collections have smaller leaves, purple pubescence, and more purple coloration in general. This is likely due to the higher amount of sunlight obtained by the epiphytic individuals. Coloration of trichomes, if variable within an individual, tends to be present in trichomes that receive more sunlight (pers. obs.).

26. Columnea ultraviolacea J. F. Smith & L. E. Skog, Novon 3: 193. 1993.—TYPE: BOLIVIA. La Paz: Prov. Sud Yungas, Huancané, 9 km from San Isidro (N of Chulumani), 2450 m, 1 Jan 1984, *Beck 8741* (holotype: LPB!; isotype: US!).

Terrestrial, herbaceous shrubs with semi-scandent stems to 1.2 m, 5–6 mm in diameter, terete, red-brown, proximally smooth and glabrous, distally pubescent; internodes 1.3–4.5 cm long; nodes and leaf scars flush with stem. Leaves in whorls of 4, isophyllous to slightly anisophyllous; laminas 2.5–4.5 cm long, 1–2 cm wide, elliptic, apex acute, base rounded to cuneate, oblique, adaxially dark green, glabrous, abaxially dark purple, strigillose, appressed-hirsute on veins with red uniseriate trichomes, lateral veins 4–5 per side, margin entire, ciliate with red uniseriate trichomes; petiole 2–6 mm long, purple, pilose to hirsute with red uniseriate trichomes. Inflorescences of a solitary flower per axil; floral bract 1, 3–8 mm long, up to 0.5 mm wide, linear, apex acute, purple, slightly pilose on outer surface, glabrous on interior, margin entire. Pedicels 8–16 mm long, purple, erect, pilose to hirsute with red uniseriate trichomes, glands near calyx 0.4–0.7 mm long, oval to elliptic, darker purple. Calyx loosely clasping corolla; lobes equal to subequal, 13–21 mm long, 1.2–2 mm wide, lanceolate to linear, acute, purple, exterior surface sparsely hirsute with red uniseriate trichomes, interior glabrous, margin entire. Corolla 4.1–4.4 cm long, 5.5–7.0 mm wide at widest point, 3.5–4.0 mm wide before limb, 2.0–2.5 mm wide at base, tubular, slightly ventricose, with a long constriction at base, constricted at limb, light yellow, exterior surface densely sericeous with uniseriate transparent trichomes, interior slightly pilose at base, glandular distally and dorsally; lobes equal to subequal, 2–3 mm long, 2–3 mm wide, semiorbicular, with purple spots on interior surface. Filaments connate at base for 8 mm, adnate to base of corolla tube for 4.5 mm, white, slightly pilose; anthers 1.5–2.5 mm long, 1.5–2.0 mm wide, quadrate, included in corolla tube. Ovary 4 mm long, sericeous with red and transparent uniseriate trichomes; style red at base and turning yellow, slightly pilose proximally, glabrous distally; stigma stomatomorphic, white, smooth, included in corolla tube. Nectary of 5 free glands. Berry 8–9 mm long, 3.0–4.5 mm wide, ovoid, dark purple with a white spot where style was attached, sericeous with red and transparent trichomes; mature seed not seen. Fig. 26.

Phenology. Flowering in January, fruiting in May.

Distribution (Fig. 12). Bolivia (La Paz); wet forests; 2400–2450 m.

ADDITIONAL SPECIMENS EXAMINED. **Bolivia.** LA PAZ: Sud Yungas, 9.2 km towards San Isidro from Huancané, *J. F. Smith & D. N. Smith 1829* (LPB, WIS).

Columnea ultraviolacea and *C. fritschii* are the only two species of *Columnea* to have leaves in a whorl of four, easily distinguishing these species from others. They are separated from each other by the yellow corollas, purple coloration, and narrow, lanceolate to linear calyx lobes of *C. ultraviolacea*. These two species are obviously sister species, based on their leaf arrangement, and are closely related to *C. moesta*, another Bolivian species, as indicated by analyses of cpDNA variation (Smith & Sytsma 1994b, c).

27. Columnea xiphoidea J. F. Smith & L. E. Skog, Novon 3: 195. 1993.—TYPE: PERU. Ucayali: Divisoria, 59 km from Tingo María on rd to Pucallpa, 17 Nov 1949–15 Jan 1950, *Allard 21230* (holotype: BH!; isotype: US!).

Herbs with stems to 2 m, 4 mm in diameter, square when dried, red-brown, proximally smooth, distally slightly to densely sericeous; internodes 1.6–4.3 cm long; nodes flush with stem; leaf scars raised. Leaves opposite, isophyllous to slightly anisophyllous; laminas 7.0–9.1 cm long, 1.4–2.0 cm wide, lanceolate, slightly falcate, apex acuminate, base rounded, oblique, adaxially green with purple margin, hirsute to pilose with uniseriate red trichomes, abaxially dull rose-red, strigose with single-celled trichomes and pilose with red uniseriate trichomes, veins slightly pilose with red uniseriate trichomes, lateral veins 6 per side, margin entire; petioles 2.5–4.0 mm long, pilose to sericeous with red uniseriate trichomes. Inflorescences each of 1–3 flowers in either axil of a leaf pair; floral bracts 1–2, 4–11 mm long, 0.5–2.0 mm wide, linear, apex acute, maroon, sericeous, margin entire. Pedicels 7–13 mm long, erect, sericeous with uniseriate transparent trichomes, eglandular. Calyx clasping corolla; lobes equal, 15–21 mm long, 2–4 mm wide, lanceolate, apex acute, exterior surface green with purple margin and some purple mottling, pilose and strigose, sericeous at base, interior bright red-purple, pilose to sericeous, margin entire. Corolla 4.2 cm long, 1.1 cm wide at widest point, 0.8 cm wide before limb, 4.5 mm wide at base, tubular, slightly ventricose, constricted at base, crimson, exterior surface villous to sericeous, more densely so towards limb, interior hirsute to pilose with glandular trichomes dorsally and distally; lobes subequal, 2 mm long, 3 mm wide, semiorbicular, with dark purple spots on interior surface. Filaments connate at base for 5 mm, adnate to base of corolla tube for 4 mm, yellow, hirsute; anthers 1.7 mm long, 1.7 mm wide, quadrate, included in corolla tube. Ovary 3 mm long, sericeous; style yellow, glabrous; stigma bilobed, yellow, papillose, included in corolla tube. Nectary of 5 free glands, the 2 dorsal glands slightly connate. Fruit and seeds not seen. Fig. 27.

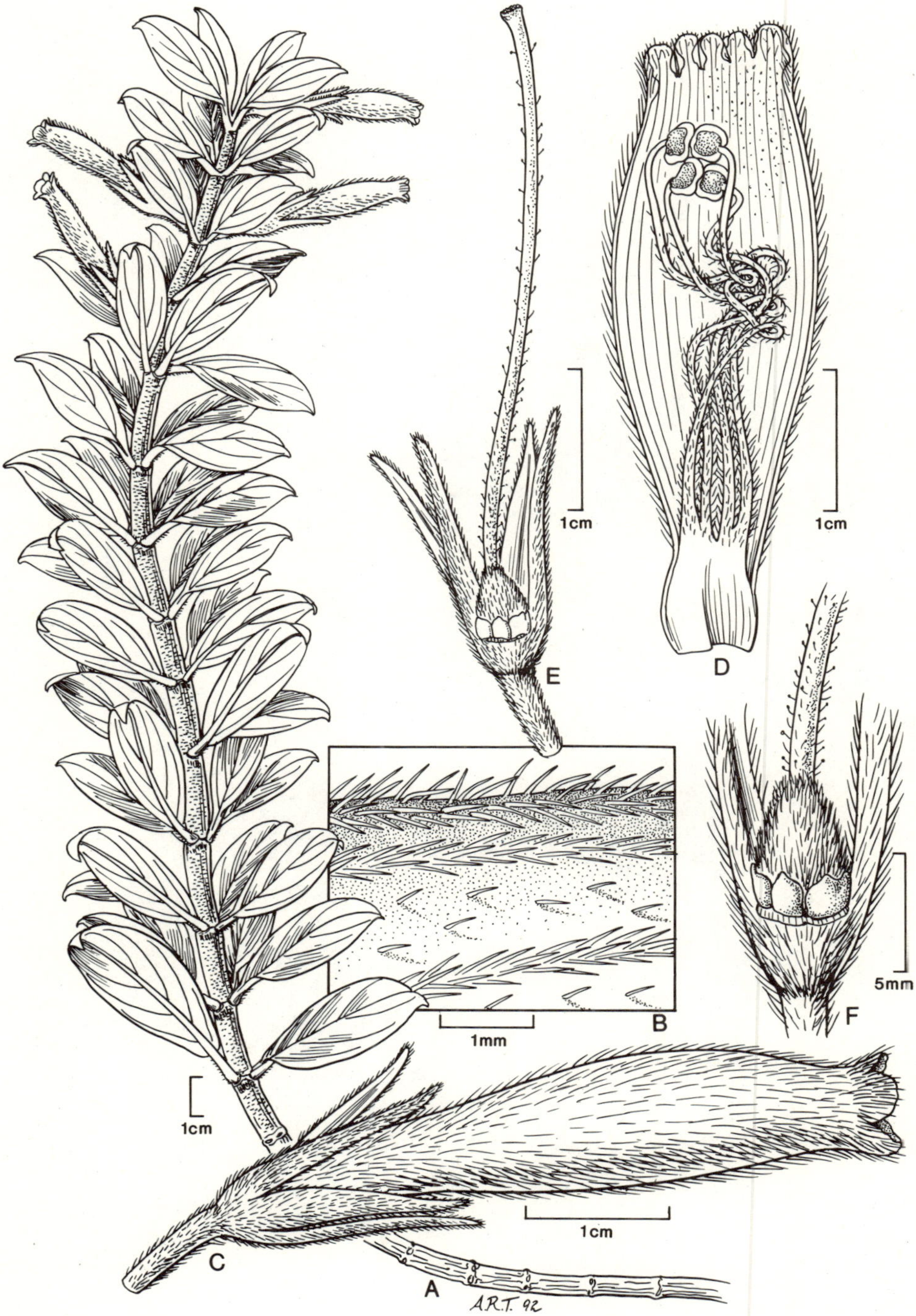

FIG. 26. *Columnea ultraviolacea*. A. Habit. B. Abaxial leaf pubescence. C. Flower. D. Corolla interior with retracted stamens. E. Gynoecium with nectaries and calyx (one calyx lobe removed). F. Base of gynoecium with nectaries (three calyx lobes removed). (Based on *Beck 8741*.)

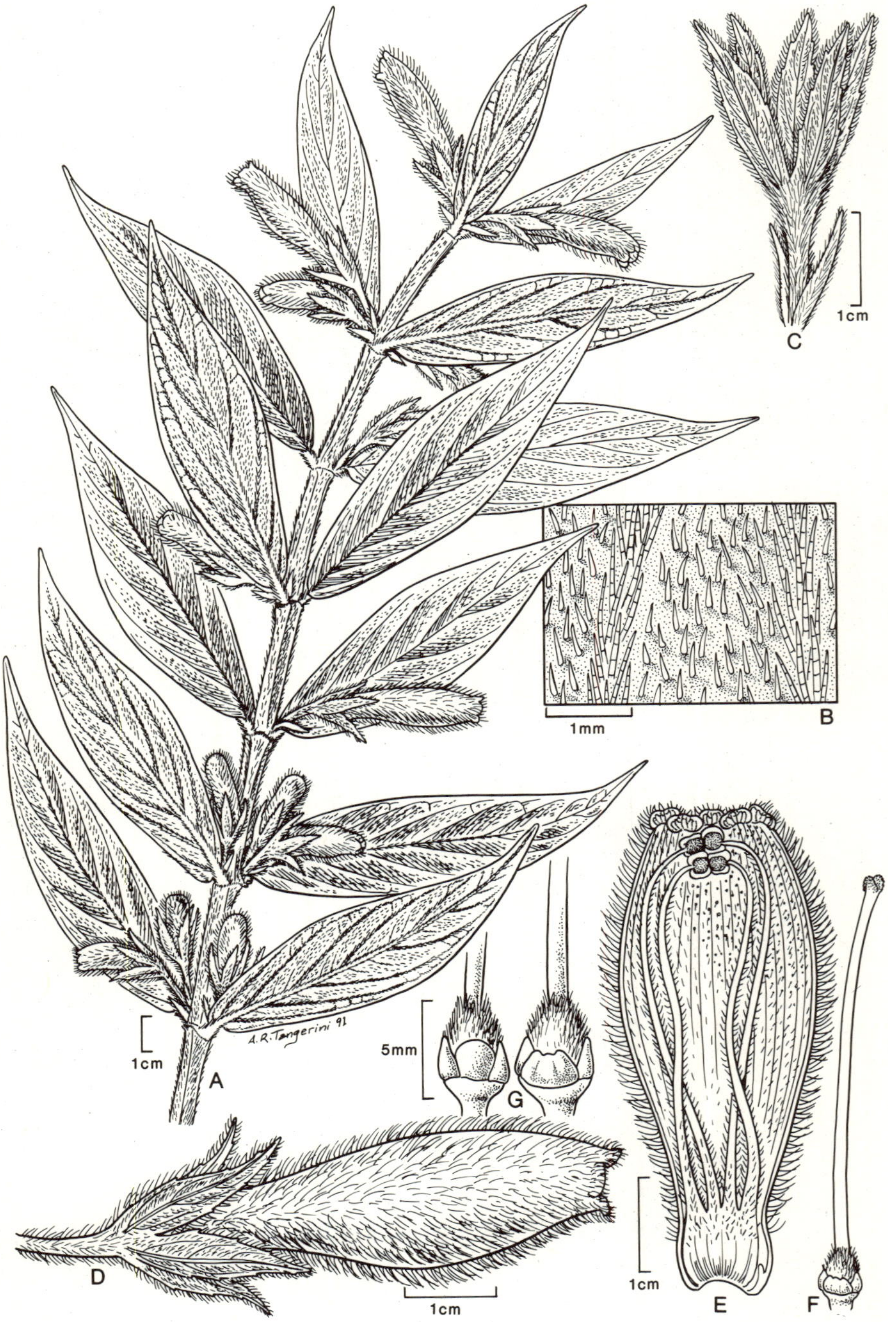

FIG. 27. *Columnea xiphoidea*. A. Habit. B. Abaxial leaf pubescence. C. Pedicel with floral bracts and calyx. D. Flower. E. Corolla interior with stamens. F. Gynoecium with nectaries. G. Bases of two gynoecia with nectaries. (Based on *Allard 21230*.)

Distribution (Fig. 8). Peru; 1600 m.

ADDITIONAL SPECIMENS EXAMINED. **Peru.** UCAYALI: La Divisoria, 59 km from Tingo María on rd to Pucallpa, *Allard 21280* (US), *Allard 21300* (US); La Divisoria, "Margarita," *Ferreyra 1012* (US).—HUÁNUCO: Cordillera Azul, ca. 39.2 km E of Tingo María on rd to Pucallpa, *Jones & Davidson 9339* (LAM); ca. 43 km E of Tingo María on rd to Pucallpa, *Jones & Davidson 9432* (LAM).

The lanceolate, isophyllous leaves and crimson, densely villous to sericeous corollas of *C. xiphoidea* readily distinguish it from any other species of *Columnea*. Although no formal cladistic analysis has been performed on *C. xiphoidea*, its morphological features suggest that it is potentially related to two different clades in sect. *Stygnanthe*. The lanceolate, slightly falcate leaves would tend to ally it with the *C. inconspicua*/*C. manabiana* clade; however, the slightly ventricose, densely sericeous to lanate corolla with dark purple spots on the interior of the lobes would place it with *C. fritschii*/*C.ultraviolacea*.

ACKNOWLEDGMENTS

I acknowledge Ken Sytsma, University of Wisconsin–Madison, for many years of support and advice. I thank Larry Skog, Smithsonian Institution, and Lars Peter Kvist, University of Aarhus, Denmark, for their contributions to this manuscript, Kandis Elliot, University of Wisconsin–Madison, and Alice Tangerini, Smithsonian Institution, for assisting and providing illustrations. I also thank Larry Skog, Paul Berry, Missouri Botanical Garden, and Christiane Anderson, University of Michigan, for their careful review and comments on the manuscript. I am grateful to the curators of the following herbaria for making their material available to me: A, B, BH, COL, CONN, CUVC, F, GH, GOET, K, LPB, M, MA, MO, MTJB, NY, P, S, SEL, U, US, W, WAG, WIS. Funding for this project was made available by NSF grant BSR–8815173 to the University of Wisconsin–Madison, and a grant from the American Gloxinia and Gesneriad Society Research Fund.

LITERATURE CITED

Beaufort-Murphy, H. T. 1983. The seed surface morphology of the Gesneriaceae utilizing the scanning electron microscope and a new system for diagnosing seed morphology. Selbyana 6: 220–442.

Bentham, G. 1876. Gesneraceae. In *Genera plantarum* by G. Bentham and J. D. Hooker, 2: 990–1025. London.

Berry, P. E. 1982. The systematics and evolution of *Fuchsia* sect. *Fuchsia* (Onagraceae). Ann. Missouri Bot. Gard. 69: 1–198.

Brown, W. R. 1938. The bearing of nectaries on the phylogeny of flowering plants. Proc. Amer. Phil. Soc. 79: 549–595.

Byrne, R., and B. Morley. 1976. Hybridization studies in *Columnea* L. (Gesneriaceae). 2. The *C. querceti* complex. Bot. J. Linn. Soc. 72: 199–210.

Ertelt, J. 1990. Growing plants in the greenhouse. Abstract from Gesneriaceae Workshop, Jardin Botanique de Montréal, Montréal, Québec, Canada.

Fritsch, K. 1893–1894. Gesneriaceae. In *Die natürlichen Pflanzenfamilien*, ed. A. Engler and K. Prantl, 4(3b): 133–185. Leipzig: Wilhelm Engelmann.

Fritze, K. J., and N. H. Williams. 1988. The taxonomic significance of pollen morphology in the *Columnea* alliance (Gesneriaceae:Gesnerioideae). Ann. Missouri Bot. Gard. 75: 168–191.

Givnish, T. J. 1984. Leaf and canopy adaptations in tropical forests. In *Physiological ecology of plants of the wet tropics*, ed. E. Medina, H. A. Mooney, and C. Vázquez-Yanes, 51–84. The Hague: W. Junk.

Grifo, F. T. 1989. Biogeography and systematics of Andean Myrtaceae with specific reference to *Myrcianthes* Berg. Amer. J. Bot. 76(6): 214. [abstract]

Hammen, T. van der. 1974. The Pleistocene changes of vegetation and climate in tropical South America. J. Biogeography. 1: 3–26.
Hanstein, J. 1865. Die Gesneraceen des Königlichen Herbariums und der Gärten zu Berlin, nebst monographischer Übersicht der Familie im Ganzen. II, 3. Linnaea 34: 225–462.
Harborne, J. B. 1967. Comparative biochemistry of the flavonoids–VI. Flavonoid patterns in the Bignoniaceae and the Gesneriaceae. Phytochemistry 6: 1643–1651.
Harling, G. 1979. The vegetation types of Ecuador–A brief survey. In *Tropical botany*, ed. K. Larsen and L. B. Holm-Nielsen, 165–174. New York: Academic Press.
Jones, C. E., and P. V. Rich. 1972. Ornithophily and extrafloral color patterns in *Columnea florida* (Morton) (Gesneriaceae). Bull. S. Calif. Acad. Sci. 7(13): 113–116.
Kiester, A. R., R. Lande, and D. W. Schemske. 1984. Models of coevolution and speciation in plants and their pollinators. Amer. Midl. Naturalist 124: 220–243.
Kvist, L. P., and J. A. Pedersen. 1986. Distribution and taxonomic implications of some phenolics in the family Gesneriaceae determined by EPR spectroscopy. Biochem. System. & Ecol. 14 : 385–405.
Kvist, L. P., and L. E. Skog. 1993. The genus *Columnea* (Gesneriaceae) in Ecuador. Allertonia 6(5): 327–400.
Linhart, Y. B., and P. Feinsinger. 1980. Plant-hummingbird interactions: effects of island size and degree of specialization on pollination. J. Ecol. 68: 745–760.
Linnaeus, C. 1753. *Species plantarum*. Stockholm.
Luteyn, J. L. 1989. Speciation and diversity of Ericaceae in neotropical montane vegetation. Amer. J. Bot. 76: 214. [abstract]
Macior, L. W. 1971. Co-evolution of plants and animals–systematic insights from plant-insect interactions. Taxon 20: 17–28.
Mansfeld, R. 1937. Gesneriaceae. In *Beiträge zur Kenntnis der Vegetation und Flora von Ecuador*, ed. F. L. E. Diels, 144–147. Stuttgart: E. Schweizerbart.
Molau, U. 1988. Scrophulariaceae–Part I. Calceolarieae. Fl. Neotrop. Monogr. 47: 1–325.
Morley, B. D. 1966. *Columnea* and aspects of its evolution. Science Notes and News, Jamaica, 2: 13–14.
———. 1971. A hybrid swarm between two hummingbird-pollinated species of *Columnea* (Gesneriaceae) in Jamaica. J. Linn. Soc., Bot. 64: 81–96.
———. 1972. Some karyotype diversity in *Columnea* L. sensu lato (Gesneriaceae). J. Linn. Soc., Bot. 65: 25–36.
———. 1973a. Ecological factors of importance to *Columnea* taxonomy. In *Taxonomy and ecology*, ed. V. Heywood, 265–281. New York: Academic Press.
———. 1973b. Materials for a treatment of *Columnea* in Panama. Ann. Missouri Bot. Gard. 60: 449–460.
———. 1974. Notes on some critical characters in *Columnea* classification. Ann. Missouri. Bot. Gard. 61: 524–525.
———. 1976a. A key, typification, and synonymy of the sections in the genus *Columnea* L. Contr. Nat. Bot. Gard. Glasnevin. 1: 1–11.
———. 1976b. Hybridization studies in *Columnea* L. (Gesneriaceae). 1. Jamaican Species. Bot. J. Linn. Soc. 72: 191–198.
Morton, C. V. 1963. A revision of *Trichantha* (Gesneriaceae). Contr. U.S. Natl. Herb. 38(1): 1–27.
Raven, P. H. 1980. Hybridization and the nature of speciation in higher plants. Canadian Bot. Assoc. Bull. Supplement to vol. 13, no. 1: 3–10.
Raven, P. H., and D. I. Axelrod. 1974. Angiosperm biogeography and past continental movements. Ann. Missouri Bot. Gard. 61: 539–673.
Rzedowski, J. 1981. *Vegetación de México*. Mexico: Editorial Limusa.
Sánchez-Burgos, A. A., and N. G. Dengler. 1988. Leaf development in isophyllous and facultatively anisophyllous species of *Pentadenia* (Gesneriaceae). Amer. J. Bot. 75: 1472–1484.
Simpson, B. B. 1975. Pleistocene changes in the flora of the high tropical Andes. Paleobiology 1: 273–294.
———. 1979. Quaternary biogeography of the high montane regions of South America. In *The South American herpetofauna: its origin, evolution, and dispersal*, ed. W. E. Duellman, 157–189. Monogr. Museum of Natural History, Univ. of Kansas. No. 7.
Skog, L. E. 1979. Family 175, Gesneriaceae. In *Flora of Panama Part IX*, ed. R. E. Woodson, Jr., and R. W. Schery. Ann. Missouri Bot. Gard. 65(3): 783–996.
Smith, J. F. 1991. The evolution and systematics of *Columnea* sections *Pentadenia* and *Stygnanthe* (Gesneriaceae). Ph.D. Dissertation. University of Wisconsin–Madison.

Smith, J. F., and K. J. Sytsma. 1994a. Evolution in the Andean epiphytic genus *Columnea* (Gesneriaceae). I. Morphological variation. Syst. Bot. 19(2): 220–235.

———. 1994b. Evolution in the Andean epiphytic genus *Columnea* (Gesneriaceae). II. Chloroplast DNA restriction site variation. Syst. Bot. 19(2): 317–336.

———. 1994c. Molecules and morphology: Congruence of data in *Columnea* (Gesneriaceae). Plant Systematics and Evolution. In press.

Stearn, W. T. 1969. The Jamaican species of *Columnea* and *Alloplectus* (Gesneriaceae). Bull. Brit. Mus. (Nat. Hist.), Bot. 4: 181–236.

Swofford, D. 1991. Phylogenetic analysis using parsimony, version 3.1. Computer program distributed by the Illinois Natural History Survey, Champaign, Illinois.

Sytsma, K. J., and J. F. Smith. 1988. DNA and morphology: Comparisons in the Onagraceae. Ann. Missouri Bot. Gard. 75: 1217–1237.

Sytsma, K. J., J. F. Smith, and P. E. Berry. 1991. The use of chloroplast DNA to assess biogeography and evolution of morphology, breeding systems, and flavonoids in *Fuchsia* sect. *Skinnera* (Onagraceae). Syst. Bot. 16: 257–269.

Vuilleumier, B. S. 1970 ["1969"]. The systematics and evolution of *Perezia* sect. *Perezia* (Compositae). Contr. Gray Herb. 199: 1–163.

Weber, A. 1973. Die Struktur der paarblütigen Partialfloreszenzen der Gesneriaceen und bestimmter Scrophulariaceen. Beitr. Biol. Pflanzen 49: 429–460.

Wiehler, H. 1973. One hundred transfers from *Alloplectus* and *Columnea* (Gesneriaceae). Phytologia 27: 309–328.

———. 1975. Name changes in Neotropical Gesneriaceae. Selbyana 1: 32–35.

———. 1977. New genera and species of Gesneriaceae from the neotropics. Selbyana 2: 67–132.

———. 1981. New species and name changes in neotropical Gesneriaceae. Selbyana 5: 378–384.

———. 1983. A synopsis of the neotropical Gesneriaceae. Selbyana 6: 1–219.

———. 1984. Miscellaneous new species in the Gesneriaceae. Selbyana 7: 328–347.

Williams, N. H. 1978. Pollen structure and the systematics of the neotropical Gesneriaceae. Selbyana 2: 310–322.

Wilson, C. L. 1974. Floral anatomy in Gesneriaceae. II. Gesnerioideae. Bot. Gaz. (Crawfordsville) 135: 257–268.

APPENDIX

Species of *Columnea* used to examine self-compatibility and interspecific hybridization under greenhouse conditions. For each cross, the female parent is listed first. After each cross, the number of attempts are listed, followed by the number of fruits produced in parentheses; the number in brackets indicates the total number of F_1 individuals from that cross. Vouchers are deposited at the herbarium cited with each collection listed. The following collections were used for crosses: *C. angustata* (*J. F. Smith 2126*, WIS), *C. colombiana* (*Wiehler 72–130*, SEL), *C. spathulata* 1 (*J. F. Smith 2229*, WIS), *C. spathulata* 2 (*Skog & Hodapp 5398*, US), *C. orientandina* (*Madison & Coleman 2537*, SEL), *C. inconspicua* (*J. F. Smith 1945*, WIS), *C. gloriosa* (*J. F. Smith 644*, WIS), *C. sanguinea* (*J. F. Smith 636*, WIS).

INTRASPECIFIC CROSSES

C. angustata × *C. angustata*, 22 (2) [55]
C. colombiana × *C. colombiana*, 53 (6) [1056]
C. spathulata 1 × *C. spathulata* 1, 5 (2) [0]
C. spathulata 2 × *C. spathulata* 2, 9 (7) [715]
C. orientandina × *C. orientandina*, 6 (2) [0]
C. inconspicua × *C. inconspicua*, 7 (2) [84]
C. gloriosa × *C. gloriosa*, 2 (1–aborted before maturity) [0]

INTERSPECIFIC CROSSES

C. angustata × *C. spathulata* 1, 1 (0) [0]
C. angustata × *C. colombiana*, 2 (0) [0]
C. angustata × *C. sanguinea*, 1 (0) [0]

C. colombiana × *C. angustata* 2 (1) [10]
C. colombiana × *C. spathulata* 1, 1 (0) [0]
C. colombiana × *C. orientandina*, 1 (0) [0]
C. colombiana × *C. gloriosa*, 1 (0) [0]
C. inconspicua × *C. spathulata* 1, 1 (0) [0]
C. inconspicua × *C. colombiana*, 3 (0) [0]
C. gloriosa × *C. colombiana* 3 (0) [0]
C. gloriosa × *C. inconspicua* 2 (0) [0]

NUMERICAL LIST OF SPECIES

1. C. atahualpae
2. C. hypocyrtantha
3. C. isernii
4. C. lophophora
5. C. nervosa
6. C. oblongifolia
7. C. poortmannii
8. C. strigosa
9. C. trollii
10. C. ambigua
11. C. angustata
12. C. antiocana
13. C. byrsina
14. C. colombiana
15. C. crassicaulis
16. C. fritschii
17. C. inconspicua
18. C. lavandulacea
19. C. manabiana
20. C. moesta
21. C. orientandina
22. C. ovatifolia
23. C. rileyi
24. C. spathulata
25. C. suffruticosa
26. C. ultraviolacea
27. C. xiphoidea

INDEX TO NUMBERED COLLECTIONS EXAMINED

The numbers in parentheses refer to the corresponding species in the text and in the Numerical List of Species above.

Acosta-Arteaga, C. E., et al. 585 (8).
Acosta-Solís, M. 5029 (8); 5455 (8); 5646 (8); 6095 (11); 6885 (24); 6904 (3); 6982 (8); 7003 (8); 7005 (8); 7143 (8); 7152 (8); 7160 (22); 7250 (8); 7404 (8); 10914 (11); 14408 (8).
Allard, H. A. 21230 (27); 21280 (27); 21300 (27).
André, E. 383 (5); 501 (5); 1371 (8); K1466 (24); K1479 (18); 2092 (8); 3526 (15); 3705 (18); 3851, June 10 (11); 3851, June 15 (17); 4688 (1); 5781 (24).
Archer, W. A. 1904 (11); 2053 (11).
Arias, H. s.n. (8).
Arteaga de García, L. A. 28 (8).
Asplund, E. 8554 (11); 9784 (8); 10080 (8); 16928 (8); 16938 (8); 17206 (8); 18932 (11); 18983 (8).
Balls, E. K. B7502 (8); B7573 (8).
Balslev, H. 1997 (13).
Balslev, H., & E. Madsen 10575 (13).
Bang, M. 401 (16); 1744 (6).
Barbour, P. 3688 (8); 4114 (8).
Barclay, A. S., et al. 3343 (8).
Barclay, H. G., & P. Juajibioy 5922 (8); 9468 (8).
Barkley, F. A., & M. Valderama 35257 (8).
Beck, S. G. 1084 (9); 8741 (26); 8772 (6).
Bell, P. R. 738 (8).
Benavides, O. 8752 (13).
Benoist, R. 2032 (8).
Bernal, R. et al. 1034 (11).
Bernardi, L. 273 (8); 579 (8); 1308 (8); 2206 (8).
Bernoullii & Cario 2010 (5).
Berry, P. E. 3419 (8).
Besse, L., et al. 81 (11); 103 (13); 180 (8); 563 (20); 564 (7); 597 (24); 621 (9); 657 (9); 1257 (11); 1272 (24); 1274 (11); 1633 (13); 1725 (4).
Boeke, J. D. 539 (8); 563 (8); 860 (8); 889 (8).
Boeke, J. D., & H. Loyola 2176 (24).
Bogner, J. 857 (8); 944 (24).
Bohlin, J. E., & M. Bohlin 943 (8).
Bohlin, J. E., et al. 1114 (8).
Breteler, F. J. 3381 (8).
Britton, N. L., & S. Brown 5415 (10).
Buchtien, O. 3902 (16).
Bunting, G. S. 2637 (24).

Heilborn, O. 477 (8); 477a (8).
Heinrichs, E. 831 (8).
Hitchcock, A. S. 21769 (11).
Holm-Nielsen, L., & S. Jeppesen 469 (11).
Holm-Nielsen, L., et al. 2679 (11); 2972 (24); 3463 (8); 5593 (8); 5992 (8); 6203 (8); 6838 (8); 17352 (8); 29241 (8); 29748 (8).
Holmgren, I. 621 (8).
Howard, R. A. 16815 (10).
Huertas G., G., & L. A. Camargo G. 5151 (8); 5884 (8); 6783 (8).
Idrobo, J. M. 2731 (8).
Idrobo, J. M., & M. Ospina H. 2411 (8).
Idrobo, J. M., et al. 3033 (8); 3450 (8).
Iltis, H. H., & M. G. Iltis E-252c (11); E-276 (24).
Iltis, H. H., et al. E-59 (24); 1055 (4); 1059 (6).
Jahn, A. 416 (8); 1031 (8).
Jameson, W. 46 (8); 146 (8); 667 (8).
Jaramillo, J. 5537 (8).
Jaramillo M., R. 6831 (8); 6909(8); 6992 (8).
Játiva, C., & C. Epling 1183 (24).
Jones, J., & C. Davidson 9339 (27); 9432 (27).
Juncosa, A. 2504 (11).
Kalbreyer, W. 1796 (8).
Killip, E. P. 3029 (8); 38096 (8).
Killip, E. P., & J. Cuatrecasas 38996 (11).
Killip, E. P., & H. García 33381 (11).
Killip, E. P., & A. C. Smith 26059 (21).
Kirkbride, J. H., & H. Chambra R. 4121 (8).
Knight, D. H. 132 (8); 515 (8); 567 (8).
Kvist, L. P., et al. 60352 (21).
Lent, R. W. 2551 (11).
Linden, J. 1454 (8).
Liebmann 9271 (5).
Liesner, R. L. 916 (11).
Lobb, W. 60 (8).
Løjtnant, B., & U. Molau 15811 (24).
Londoño, X. 240 (13); 406 (8).
López, A., & A. Sagástegui 5389 (8).
López, A., et al. 6607 (8).
López Figueiras, M. 8335 (8).
López P., S. 4194 (8); 4219 (8); 4242 (8).
Lozano C., G. 3350 (8); 3521 (8); 4044 (8).
Lozano C., G., et al. 819 (8); 2557 (8); 4498 (8).
Luer, C. A., et al. 809 (8); 3177 (13); 3535 (20); 4496 (13); 4717 (15); 4877 (2); 4890 (20); 4936 (21); 4957a (9).
Lugo, H. 34 (11); 230 (11); 468 (11); 676 (11); 757 (8); 1111 (11); 1319 (8); 1805 (11); 1848 (8); 1912 (8); 4506 (11); 5702 (11).
Luteyn, J. L., & Giraldo 12644 (25).
Luteyn, J. L., & M. Lebrón-Luteyn 5607 (8); 5820 (11).
Luteyn, J. L., et al. 4970 (8); 6010 (8); 8803 (8); 8830 (23); 12288 (25).
Luther, H., et al. 887 (13).
Maas, P. J. M., & R. Jaramillo M. 1784 (8).
Maas, P. J. M., & T. Plowman 1838 (13).
MacBryde, B. 99 (8); 123 (8).
MacBryde, B., & J. D. Dwyer 1275 (8).
MacDougall, T. 466 (5).
Madison, M. T. 3873 (13); 4004 (24); 4029 (24).
Madison, M. T., & L. Besse 7115 (13).
Madison, M. T., & F. R. Coleman 2537 (21).
Madison, M. T., et al. 3192 (11); 3322 (11); 3370 (17); 3420 (21); 4419 (13); 4451 (13); 4991 (24).
Mason, H. L. 13926 (8).
Mathias, M. E., & D. Taylor 5180 (24).
Matuda, E. 1626 (5); 1860 (5); 1870 (5).
Melampy, M. 381 (8); 831 (8).
Mexia, Y. 7615 (8).
Montenegro, E., & S. Espinal 5A (8).
Moore, H. E. 7773 (11).
Mora, L. E. 1376 (8); 2112 (8).
Mori, S., & J. Kallunki 5471 (11).
Neill, D., et al. 5790 (11).
Núñez V., P., & E. Bengoa 8551 (3).
Núñez V., P., & F. Motocanchi 8808 (20).
Øllgaard, B., & H. Balslev 9506 (8).
Orozco, C. I., et al. 618 (11).
Otero, J. I. 281 (10).
Overton, D. W. 0-56-100 (8).
Pachano, A. 180 (8).
Padilla, I. 1082 (8).
Pennell, F. W. 2423 (8); 6947 (8); 9384 (8); 14038 (20).
Pennell, F. W., & E. P. Killip 6608 (8); 6629 (8).
Penland, C. W., & R. H. Summers 145 (11); 415 (8); 572 (8).
Pérez A., E., & J. Cuatrecasas 5266 (8); 6151 (8).
Peyton, B., & S. King 1405 (6).
Peyton, B., & S. Tilney Peyton 1382 (6); 1488 (6).
Pfeifer H. W., et al. 2836 (10); 2471 (10); 2860 (10); 2603 (10); 3014 (10).
Plowman, T., & E. W. Davis 4128 (8); 4312 (8).
Plowman, T., et al. 13411 (24).
Poeppig, E. D1626 (20).
Poortmann, H. 207 (5); 243 (5); 265 (7); 501 (5).
Ramos et al. 1357 (25).
Rangel, O. 2557 (8).
Reichenbach, fil. 63 (8); 65 (8); 72 (8); 73 (8).
Rimbach, A. 104 (8); 176 (8); 244 (8); 265 (8); 281 (8); 493 (8); 648 (8).
Rivet, M. (24).
Robinson, J. W. L. (13).
Roe, K., et al. 723 (5).
Rooden, J. van, et al. 449 (11).
Rose, J. N., & G. Rose 22281 (4); 23461 (1).

INDEX TO SCIENTIFIC NAMES

Accepted names are in roman type; the main entry for each is in **boldface**. Synonyms are in *italics*.